Efficient Comfort Conditioning

The Heating and
Cooling of Buildings

AAAS Selected Symposia Series

Published by Westview Press
5500 Central Avenue, Boulder, Colorado

for the

American Association for the Advancement of Science
1776 Massachusetts Ave., N.W., Washington, D.C.

Efficient Comfort Conditioning

The Heating and Cooling of Buildings

Edited by
Walter G. Berl and W. Richard Powell

AAAS Selected Symposium **27**

AAAS Selected Symposia Series

Copyright © 1979 by American Association for the
Advancement of Science

Published in 1979 in the United States of America by
  Westview Press, Inc.
  5500 Central Avenue
  Boulder, Colorado 80301
  Frederick A. Praeger, Publisher

Library of Congress Catalog Card Number: 79-2046
ISBN: 0-89158-290-8

Printed and bound in the United States of America

About the Book

This timely study deals with the heating and cooling of
buildings using innovative systems that can reduce fossil
fuel and electric energy requirements by as much as 80 percent.
Emphasis is placed on thermal storage, utility rate structures, peak load problems, and cogeneration of heat and power
in small-scale applications. The first several chapters
treat promises and problems of solar energy use for efficient
comfort conditioning. Other contributions deal with the
social implications of future energy efficiency requirements
with a focus on the community.

About the Series

The *AAAS Selected Symposia Series* was begun in 1977 to
provide a means for more permanently recording and more
widely disseminating some of the valuable material which is
discussed at the AAAS Annual National Meetings. The volumes
in this *Series* are based on symposia held at the Meetings
which address topics of current and continuing significance,
both within and among the sciences, and in the areas in which
science and technology impact on public policy. The *Series*
format is designed to provide for rapid dissemination of
information, so the papers are not typeset but are reproduced
directly from the camera-copy submitted by the authors, without copy editing. The papers are organized and edited by
the symposium arrangers who then become the editors of the
various volumes. Most papers published in this *Series* are
original contributions which have not been previously published, although in some cases additional papers from other
sources have been added by an editor to provide a more comprehensive view of a particular topic. Symposia may be reports of new research or reviews of established work, particularly work of an interdisciplinary nature, since the AAAS
Annual Meetings typically embrace the full range of the
sciences and their societal implications.

WILLIAM D. CAREY
Executive Officer
American Association for
the Advancement of Science

Contents

List of Figures

List of Tables

About the Editors and Authors

*Walter G. Berl, supervisor of the Fire Problems Group in
the Johns Hopkins University Applied Physics Laboratory, is
a physical chemist. A fellow of AAAS and the New York Academy
of Sciences, he has served as editor of several professional
journals and of the 9th, 10th, and 11th International Symposia
on Combustion and of* Physical Methods in Chemical Analysis
(4 Vols.; Academic Press, 1950-60).

*W. Richard Powell, senior staff physicist in the Space
Department at the Johns Hopkins University Applied Physics
Laboratory, has worked extensively on energy systems and
biomedical devices. His publications include "Absorber for
Solar Power"* (Applied Optics, 13: 2430).

*Joseph G. Asbury is manager of systems evaluation for the
Argonne Storage Program in the Energy and Environmental Systems Division of Argonne National Laboratory. His specific
areas of specialization are economics and physics, and he has
collaborated with Caruso, Giese and Mueller in the assessment
of technical and economic impacts of a wide range of space
heating and cooling systems, with particular emphasis on
electric and electric-assisted devices. Recent papers by this
group have presented a detailed assessment of long-run supply
benefits of electric storage heating and storage air-conditioning for utility load leveling; a general comparative
systems analysis of the interface of solar systems with the
electric utility supply network; and the first general treatment of the effects of temperature stratification on solar-
thermal system performance.*

*John C. Bell, a contractor with the Office of Technology
Assessment's Energy Program, deals with system/policy and
economic/financial analyses. He was formerly the liason
between MIT's Energy Laboratory and the Office of Energy R&D
Policy at the National Science Foundation, and he has publish-*

ed articles on inventory influence on company performance, two-dimensional submerged jet flow, cost/benefit analysis of synthetic fuels, and solar energy.

Gerald E. Bennington, associate department head of the Advanced Energy and Resource Systems Analysis Department at The MITRE Corporation, specializes in solar energy. He is president of the Northern Virginia Chapter of the Virginia Solar Energy Association, an associate professorial lecturer in operations research at The George Washington University, and a referee for numerous professional journals. His publications include "Solar Energy: A Comparative Analysis to the Year 2000" (MITRE, 1978) and "An Economic Analysis of Solar Hot Water and Space Heating" (MITRE, 1976).

Karl W. Böer, professor of physics and engineering at the College of Engineering, University of Delaware, has conducted extensive research on solid state physics, solar energy conversion, photovoltaics, and solar heating of buildings. He is director of International Solar Energy Society, chairman of the board of SES, Inc., a member of the Executive Committee and Energy Technology Group of the American Electrochemical Society, a fellow of the American Physical Society, and a senior member of IEEE. He is the author of numerous publications and holds many patents in his field.

Joseph V. Caruso is an assistant economist in the Energy and Environmental Systems Division at the Argonne National Laboratory. His specific area of specialization is economic analysis of alternative energy supply technologies, and he has recently collaborated with Asbury, Giese and Mueller on the assessment of technical and economic impacts of electric, electric-assisted, and other space heating and cooling systems.

David Claridge, an analyst with the Energy Group at the Office of Technology Assessment, is currently working on energy conservation and solar technologies. He was a AAAS Congressional Science Fellow in 1976, and his publications include articles on superconducting Josephson junctions, passive solar heat gain, window management strategies, and improved window materials. He is coauthor of Applications of Solar Technology to Today's Energy Needs *(2 Vols.; OTA, 1978).*

Donald W. Connor is senior physicist in the Division of Environmental Impact Studies at the Argonne National Laboratory. His specific area of specialization is environmental/ economic analysis. He has recently published a paper on low temperature storage in Handbook of Solar Energy Technology *(Marcel Dekker, in press).*

Robert Thomas Crow, program manager for demand and conservation in the Energy Analysis Department of the Electric Power Research Institute, has been studying energy consumption behavior and the impacts of conservation. He has published articles on the economics of demand and conservation, the market for electric automobiles, evaluation of heat pumps, technological change and energy conservation, and the demand for new goods.

John Furber, an analyst with the Energy Group at the Office of Technology Assessment, has studied energy and environmental issues, politics, and economics, and has focused on solar energy. He was a participant in the US-USSR Joint Solar Energy Workshop in 1977 and has presented several papers on solar concentrator/photovoltaic power systems. He is co-author of **Applications of Solar Technology to Today's Energy Needs** (OTA, 1978) and photographer/illustrator of **Sun and Mankind** (Moscow: International Library, "Detskaya Literatura," 1979).

Robert F. Giese is an environmental engineer in the Energy and Environmental Systems Division of Argonne National Laboratory. His specific area of specialization is analysis of storage systems interfacing with electric utilities. He has held Woodrow Wilson and Alfred P. Sloan Fellowships, and he has recently collaborated with Asbury, Caruso, and Mueller on assessment of technical and economic impacts of space heating and cooling systems, with particular emphasis on electric and electric-assisted devices.

John Karkheck is a physicist with the Department of Energy and Environment at Brookhaven National Laboratory. His current area of interest is the application of district heating systems to US urban areas. He has presented and published papers on district heating and on waste heat management and utilization at various professional meetings and in **Science** magazine.

Henry Kelly, project leader with the Solar Energy Project at the Office of Technology Assessment, was a AAAS Congressional Science Fellow in 1974-75 and received a distinguished service award from ACDA. He has published papers on the physics of light scattering, arms control and disarmament, energy policy, and solar energy technology.

Gerald S. Leighton is assistant director for communities and buildings energy systems in the Division of Buildings and Community Systems at the Department of Energy. He specializes in engineering administration, was formerly director of the Integrated Utility Systems Program at HUD, and was president

*of the International Energy Agency in 1975. An active member
of the American Society of Heating, Refrigerating and Air-
conditioning Engineers, he has published some 30 papers,
including "Economic Guidelines for Total Energy Systems"
(*ASHRAE Transactions, *Vol. 80, Pt. 2, 1974), which won the
ASHRAE "Best State of the Art Technical Paper" award in 1975.*

*Ronald O. Mueller, assistant physicist in the Energy and
Environmental Systems Division of Argonne National Laboratory,
works in the area of energy economics and energy technology
assessment. He, Asbury, Caruso, and Giese have collaborated
on assessments of technical and economic impacts of space
heating and cooling systems, particularly electric and electric-assisted devices.*

*James Powell, head of the Fusion Technology Group at
Brookhaven National Laboratory, is a nuclear engineer. A
member of several professional societies, he has published
extensively on district heat technology; aspects of fusion
technology and synthetic fuel production; and applications
of superconductivity, including magnetic levitation and power
transmission.*

*Jerome H. Rothenberg, manager of the Utility and Energy
Systems Division of the Office of Policy Development and Research at HUD, is a specialist on utilities and energy technology and is concerned with international technical exchanges
and program development and management. A registered professional engineer who holds several patents, he is the author
of a paper on urban planning for arid zones as well as papers
on the MIUS program.*

*Peter C. Spewak, group leader with the Advanced Energy
and Resource Systems Analysis Department at The MITRE Corporation, is working on solar energy commercialization. His publications include "State-of-the-Art of Solar Heating and
Cooling in the U.S.," "A Comparative Analysis of Solar Energy
Applications to the Year 2000 and Beyond," and "Systems De-
scriptions and Engineering Costs for Solar-Related Technologies, Volume II: Solar Heating and Cooling of Buildings" (All
published by the MITRE Corporation, 1977).*

*Robert G. Uhler, director of the Energy Analysis Depart-
ment of the Electric Power Research Institute, specializes in
energy economics. He is the author of* Rate Design and Load
Control: Issues and Directions *(EPRI, 1977).*

Efficient Comfort Conditioning

The Heating and
Cooling of Buildings

Introduction

Walter G. Berl

The objective of this book is to present a
status report on innovative work in one of the oldest
technologies and one of mankind's crucial life support
systems: the efficient heating and cooling of buildings.

At present, in all the so-called developed countries,
this task - an adjustment by a few degrees of the "inside"
temperature above or below that of the "outside" - is done
by the convenient but far from efficient burning of fossil
or nuclear fuels either directly to supply heat or indirectly
in power stations to supply electricity for cooling devices
or electric heaters.

How large these fuel needs are is shown by the fact
that end-use enthalpy needs for this heating and cooling by
less than 100°C require 34% of all the U.S. energy consumption. In New Jersey and Delaware, for example, one-
third of all liquid fuel and two-thirds of all natural gas
is consumed for the purpose of space and water heating.

The current prodigious energy consumption along traditional lines cannot continue for long. Unusually richly
blessed as we have been in this country with indigenous
energy supplies (first wood, then coal, now oil and gas),
the exhaustion of the latter two is in sight, even though
the Alaska and off-shore deposits will extend the time to
exhaustion somewhat. The competition for the remaining
resources of fossil fuels around the world is so keen that a
substantial rise in absolute cost is unavoidable. Prospects
for 'unlimited' cheap energy, promised at the beginning of
the nuclear age, have become dim, indeed. The complexity of
nuclear power is now apparent. Even the fall-back on coal
is in question due to complications with carbon dioxide and
other emissions and the land use and water limitations of
coal mining and conversion.

1

No wonder that the disappearance of cheap fossil fuels
(and the impending demise of fossil fuels altogether)
necessitates a reassessment of this enormous energy sink.
Even today it is not fully recognized that houses, too, can
be 'fuel guzzlers' and that, in the future, much more attention must be given to their energy effectiveness. Energy
"tuning" of houses is a virtually unknown occupation.

The thrust of the discussion is that, without changes
in the prevailing living standards, very substantial reductions in energy needs can be made and that the lockstep
dependence on fossil or nuclear fuels can be broken. In so
doing, we live up to the "new ethics" that requires

1.   Non-renewable materials (including fossil fuels)
      must be used only if they are indispensible, and
      then only with the greatest care and the most
      meticulous concern for their conservation.

2.   The aim should be to obtain the maximum of well-
      being with the minimum of consumption: to be
      elegantly economical, never crudely wasteful.

We discuss elegantly frugal ways of providing heating
and cooling needs, largely by schemes that use renewable
energy resources and by methods that are suited for the task
at hand. We also deal with the not-so-trivial institutional
hurdles that have to be recognized and overcome so that
these schemes can be brought into use.

Community Energy Systems
The Promise of Tomorrow

Gerald S. Leighton

ABSTRACT

Communities, now experiencing rising energy costs, are
faced with providing community services in a manner that
maintains a high standard of living, is environmentally
acceptable, and provides consumers with price-stable,
reliable services. The objective is to do this and con-
serve energy. Two elements of community energy systems
which have significant energy-saving potential are cogeneration and land use planning. Cogeneration is broadly
defined as the joint production of more than one energy
good. Key to the development of cogeneration community
energy systems is the element of community design and land-
use planning, which allows for the optimal spatial arrangements of facilities to reap the benefits of integrated
systems.

Program activities of Federal agencies engaged in these
efforts are discussed, including the programs of the Department of Housing and Urban Development, the Council on
Environmental Quality, the National Science Foundation, and
the Department of Energy. The initial comprehensive efforts
at DOE are directed at the Grid-Connected Integrated
Community Energy System. In this system the electric utility
partnership alleviates some of the initial restrictions
concerning franchise arrangements. It further allows for
the maximization of energy use efficiency by allowing for
thermal demand control with unrestricted electricity inter-
change between the system and the utility network. The
system is designed to serve communities with heating,
cooling, and electricity. It is well along toward initial
operation in the early 1980's.

Figure 1. Community Energy Supply and Demand

The Need for Integrated Community Energy Systems

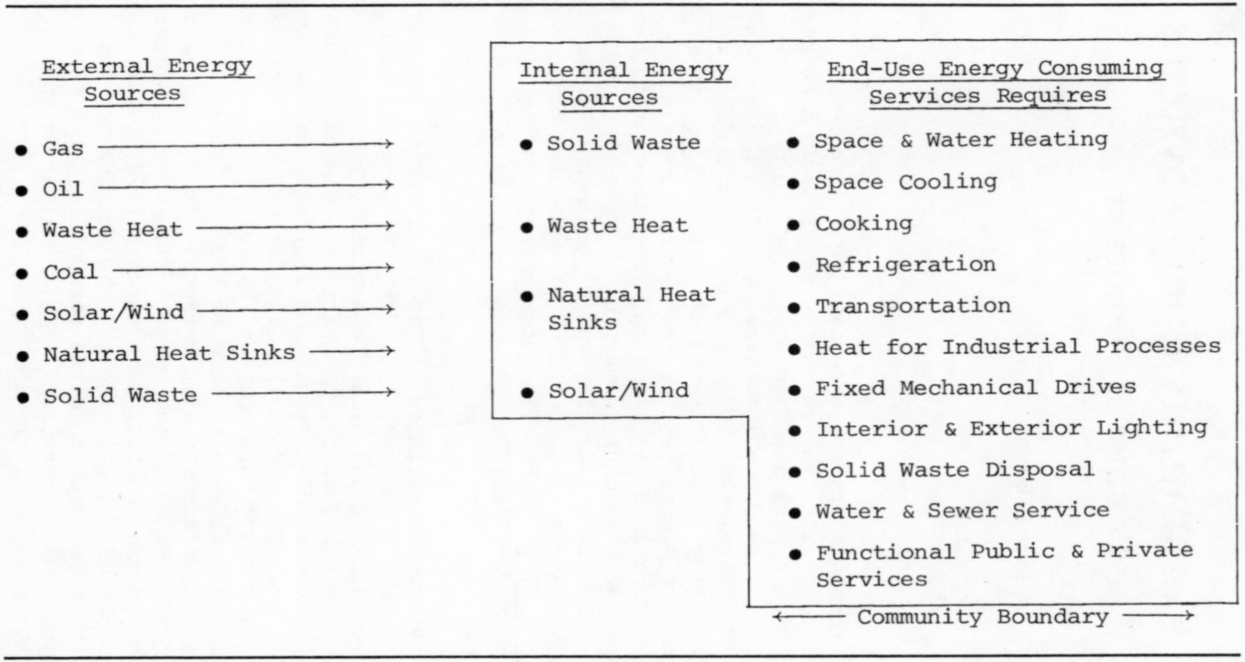

External Energy
Sources

- Gas ⟶
- Oil ⟶
- Waste Heat ⟶
- Coal ⟶
- Solar/Wind ⟶
- Natural Heat Sinks ⟶
- Solid Waste ⟶

Internal Energy
Sources

- Solid Waste
- Waste Heat
- Natural Heat Sinks
- Solar/Wind

End-Use Energy Consuming
Services Requires

- Space & Water Heating
- Space Cooling
- Cooking
- Refrigeration
- Transportation
- Heat for Industrial Processes
- Fixed Mechanical Drives
- Interior & Exterior Lighting
- Solid Waste Disposal
- Water & Sewer Service
- Functional Public & Private
  Services

⟵ Community Boundary ⟶

1. Introduction

Events of recent years have created public demand at the community level for energy supply systems that are energy-conserving, safe, environmentally acceptable, reliable and "price stable" - that is, consumers can expect that their energy expenditures will be a relatively constant share of their total budgets. The Integrated Community Energy Systems program at the Department of Energy is developing community-scale energy systems with these characteristics. These systems will represent an integration of community design planning and energy technology concepts and will help achieve the national goal of conserving energy and, in particular, of conserving scarce fuels. This paper presents the Department of Energy's concept of Integrated Community Energy Systems, and the Federal Government's role in their development and commercialization.

2. Community Energy Systems Concept

Communities consume large quantities of energy in the operation of buildings and facilities which support the activities and functions that constitute a community. This environment, while consuming energy and materials, generates waste products - solid and liquid wastes, rejected heat, etc. - which, if recovered and reused within the community, can become an energy resource which is an economic asset rather than a community liability. The range of energy sources and energy-consuming services at the community level is shown in Figure 1. The full utilization of energy resources, both natural and waste streams, requires an integrated approach to the development and operation of the energy systems that transmit and distribute energy in the community environment. This integrated approach is that of Community Energy Systems. Realizing that energy saved is more valuable than the equivalent of new energy resources, Community Energy Systems can be designed to incorporate all of the community energy resources into an efficient system providing energy services to the community.

Integrated Community Energy Systems (ICES) coordinate various energy services, such as electricity, cooling and heating, hot water, solid and liquid waste treatment, and others in such a way that the energy normally wasted in producing a service is used to fuel other services. For example about two-thirds of the energy used in electrical generation and distribution is wasted. That is, only one-third of the energy input is transformed into a useful service. Integrated systems provide for siting of the facility within the community thereby facilitating the incorporation of waste

Table 1. Partial List of Promising Technologies and Concepts

Prime Movers

    Diesel Engines
    Gas Turbines
    Rankine Engines
    Stirling Engines
    Windmills
    Combined Cycles
    Fuel Cells

Energy Conversion

    Solar Energy Conversion
    Coal Gasification

Storage

    Advanced Batteries
    Thermal Storage Systems
    Nonconvective Ponds

Fuel/Waste Processing

    Anaerobic Digestion
    Pyrolysis
    Coal Gasification

Liquid Waste Treatment Effluent Disposal

    Advanced Physical/Chemical Processes
    Thermal Enhancement of Effluent Processing

Distribution Infrastructure

    Advanced Thermal Distribution Techniques
    Waste Collection Systems
    Control Systems

Environmental Control Technologies

    Cooling Towers
    Combined Stack Gas Scrubbing and Waste
        Water Purification

End-Use Technologies/Climate Control

    Solar Heating and Cooling
    Heat Pumps
    Compression/Absorption Chillers
    Etc.

heat recovery into the system. The efficiency of these integrated systems ranges to as high as 85 percent. In addition, Community Energy Systems offer considerable potential for fuel substitution, thereby allowing the use of nonscarce fuel resources (such as coal) which would not be economically usable in smaller unintegrated systems. The overall energy efficiency of these systems is maximized by integrating the energy system with community or building design considerations such as layout and composition.

Some of the promising technologies and concepts for various integrated systems are listed in Table 1. It should be emphasized that the ICES concept is more than a hardware system - although equipment certainly is a part of an ICES. The concept also includes ways of designing and arranging community structures to minimize energy consumption. Most importantly, an Integrated Community Energy System is a comprehensive energy management concept applied at the community level that seeks an optimal combination of both of these dimensions to meet the energy requirements of a particular community in an energy-conserving, safe and reliable, economically stable, and environmentally acceptable fashion. One may refer to the concept as the integration of energy requirements, the sciences of human settlements (ekistics), environmental considerations, and economics. This integration is referred to as the "4-E's" (Energy, Ekistics, Environment and Economics).

The potential energy benefits of ICES are very large. ICES can be from 50 to 100 percent more energy efficient than conventional separate single utility supplies. With the application of ICES and community designs that are possible by the year 2000, 50 percent of the energy required by current designs could be saved in each new application. In addition to absolute energy savings, these systems have the advantage of allowing significant shifts from scarce to more abundant or renewable fuels. The intermediate scale technologies involved can utilize coal, urban wastes, biomass, solar and geothermal inputs in place of scarce gas and oil.

The definition of "community" as used in the phrase "Integrated Community Energy Systems" is "a complex of facilities (and open spaces) employed in human activities and connected by networks for moving people, messages, goods and services in the residential, commercial, industrial, agricultural, recreational or institutional sectors." The networks may be transportation routes or modes, pipelines, communications links, telephone or electrical transmission lines, etc. Superimposed on the physical landscape of any

Table 2. Preliminary Classification of Target Communities

Residential

  Single-Family Homes
  Townhouse Complex/Low-Rise Housing
  High-Density Urban
  Mixed-Density Communities

Institutional

  University/College Campus
  Hospital/Medical Complex
  Government Complex
  Research Park
  Military Base
  Correctional Facility

Mixed Land Use Communities

  Residential/Commercial
  Residential/Commercial in Support of One or More
    Central Institutions
  Residential/Commercial in Support of an Industrial
    Park
  New Communities
  New-Town-In-Town
  Small Town
  Planned Unit Development (PUD)

Commercial

  Central Business District
  Shopping Complex - District, Regional
  Office Complex
  Sports Complex
  Recreation/Amusement Facility
  Airport

Other Major Determinants of Community Characteristics

New Planned Development
Redevelopment
Retrofit
Size (Population, Floor Area, Land Area)
Building Construction
Climate
Geological Characteristics (Availability of Resources, Groundwater, etc.)

community are complex political, social and economic systems,
and jurisdictions that determine the types and levels of the
community's activities.   Thus, a "community" may be as diverse
as a municipal or suburban business district, a farm area, or
a multizoned planned unit development, to name only a few.
Furthermore, the ICES Program is conceived for communities
in various stages of development in both new and redeveloped
areas. A preliminary classification of target communities
is presented in Table 2.

3. Barriers

Numerous barriers currently exist which impede the development and implementation of integrated community energy
systems. The barriers to implementing systems utilizing
waste heat (cogeneration, district heating and cooling, total
energy systems) appear to be primarily economic and institutional. District heating requires pipe networks, the high
cost of which may make the price of the heat noncompetitive.
Other factors include possible siting restrictions, and
the general trend in the past to promote large, remotely
located energy facilities. New architectural approaches for
utilizing above-ground systems have to be investigated and
evaluated. There are several institutional problems. Many
utility companies are prohibited from engaging in providing
services other than electricity to customers. Another
regulatory/economic consideration is the ability to lock a
customer into utilizing the district heating service for a
time span that allows for the amortization of the investment. Without such a commitment, a utility company, be it
the electric utility or a district heating company, may find
itself having made a large investment in the system and not
having a continuing source of revenue to meet the debt service.

Before the potential for converting urban wastes to
beneficial energy supplies can be realized, existing barriers
must be addressed. These barriers are numerous. Typically
the retrofit installation of waste-to-energy systems requires
significant capital outlays, and although they appear to be
economically feasible on a total or life-cycle basis, muni-
cipalities are cash poor up front. Also, cities and large
utilities have generally been the partners in waste-to-energy
systems and most systems have been quite large - 1000 tons
per day is the norm. The development of smaller scale systems is needed to open up the potential in small communities,
and also to provide flexible satellite type systems for urban
areas; that is, to match waste streams and end products to
optimal end uses. One of the most serious institutional
problems is posed by the interface between those organiza-

tions charged with the collection and disposal of wastes, generally a city sanitation department, and those involved with energy production. These must be overcome on a site or political subdivision specific basis, as well as on a national and state basis since waste collection and disposal and energy production are regulated by separate entities at all levels of government. European experience indicates that wastes can be put to productive use in providing energy when the requirements to dispose of wastes and the requirements to produce energy are closely integrated. The key is to find methods to achieve this integration in the free enterprise structure of the United States.

4.   The Federal Role

The paramount role in solving the country's energy problems belongs to the private sector. The Federal Government is assisting the private sector by setting national direction and priorities. The President is attempting to do this in the National Energy Plan by stimulating private sector efforts, removing existing technical, institutional, economic, environmental or other barriers, and by conducting the appropriate research, development and demonstrations. All of these strategies are aimed at reducing the risks associated with the development and commercialization of energy-efficient integrated systems, thereby accelerating their adoption by the private sector. Governmental assumption of responsibility for developing more advanced and higher risk integrated technologies and community design concepts to the state of commercial feasibility is therefore required. Federal initiatives are also necessary to ensure that community energy conservation measures will reflect national issues and priorities as well as the more limited goals of private or local interests. The balance between public and private commitment to community energy conservation will be established through a rational assessment of the near- and long-term economic, social, and environmental costs and benefits of alternative conservation technologies, policies, and programs.

5.   Policy

The President's National Energy Plan contains proposals which, directly and indirectly, will promote the development and commercialization of community energy systems. The Plan calls for changes in public utility regulatory policies; incentives for the expanded use of coal, solar and wind-generated, and advanced energy resources; incentives for the cogeneration of electrical and other forms of useful energy; prohibitions against the use of natural gas and petroleum; and business tax credits promoting the use of such energy resources and systems.

Under the Plan, facilities generating electricity would be assured of receiving fair rates from utilities to sell surplus power and buy stand-by power. Cogeneration facilities would be exempted in whole or part from State and Federal public utility regulation. In addition, they would be entitled to use public utility transmission facilities. The Plan proposes an additional 10 percent tax credit (above the existing investment tax credit) for industrial and utility cogeneration equipment, to encourage the conservation of oil and gas. A similar provision is aimed at encouraging conversion from oil and gas to other energy resources.

Within this general framework, various emphases are possible. The rate at which integrated community energy systems penetrate the market will depend upon the final mix of investment tax credits, regulations, Federally sponsored R&D, fuel allocation priorities, conversion requirements, etc.

6. DOE ICES Approach

The comprehensive integrated approach to community energy conservation taken by DOE is to utilize alternative energy resources and select a combination of technologies, community designs and governmental arrangements that will provide a highly efficient integrated community energy system. This "systems" approach combines the physical and institutional analysis and design of communities with the development and adoption of innovative and/or advanced technology integrated community systems to minimize consumption of nonrenewable energy resources in providing a community's combined requirements for energy-consuming services. Input energy sources for integrated systems may include low-grade waste-heat recovery, solid and liquid waste-heat utilization, solar, geothermal, and seawater heat dissipation, energy, and less scarce fuels such as coal and biomass.

This approach addresses three closely coordinated areas of effort:

● Integrated Systems - to increase the efficiency for the production and distribution of energy services, develop integrated community energy systems that meet targeted energy efficiency, scarce fuel savings, and other performance criteria (safety, reliability, maintenance, operation, environmental control) at minimum cost.

● Community Design - to reduce the demand for energy ser-

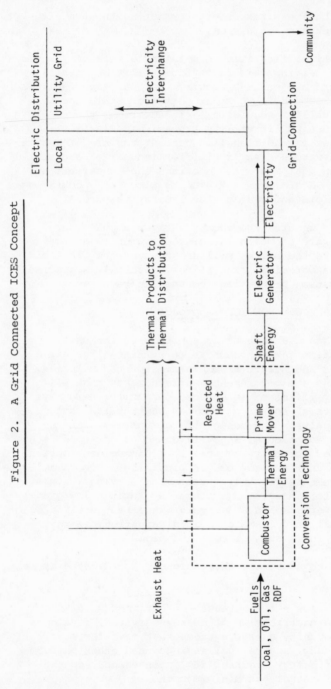

Figure 2. A Grid Connected ICES Concept

vices, develop community designs that meet targeted energy
performance criteria at minimum cost through trade-offs
in building design, community layout, activity mix, and
density or other factors of development choice.

• Implementation Mechanisms - to develop mechanisms for
implementing integrated community energy systems that
achieve maximum energy conservation by combining the
design of communities and the design of energy systems
through a coordinated approach to community energy
management that integrates both the community development
process and the energy system development process.

The development and commercialization of integrated
systems is facilitated by involvement of the principal imple-
menters from the outset of programs. We have been working
and will continue to work with individual building owners,
developers, utilities, equipment manufacturers, other Federal agencies, State and local governments, and professional
and trade associations. Simultaneous consideration of
institutional and financial barriers that inhibit the acceptance of integrated systems is undertaken so that information will be available to allow governments at all levels
to address and remove these barriers as needed.

7.   DOE ICES Programs

Integrated Systems

Two very important concepts of new ICES demonstrations -
the Grid-Connected ICES and the Coal-Using ICES - are well
along towards initial operation in the early 1980's.

The Grid-Connected ICES Program objectives are to develop and demonstrate grid-connected systems using current
technology to serve communities with heating, cooling, and
electricity. The Grid-Connected ICES is a concept (depicted
in Figure 2) which avoids the electric-power/thermal-demand
imbalance by:

(1) Operating at a level consistent with the thermal demand
     of the community; and

(2) (a) Supplying the electrical energy generated (now
          considered the "by-product") directly to the grid
          of the local electric utility which serves the
          region within which the community is situated, or

     (b) Supplying the electrical energy generated to an
          electrical grid for the community with the local

Figure 3. Grid Connected ICES Program

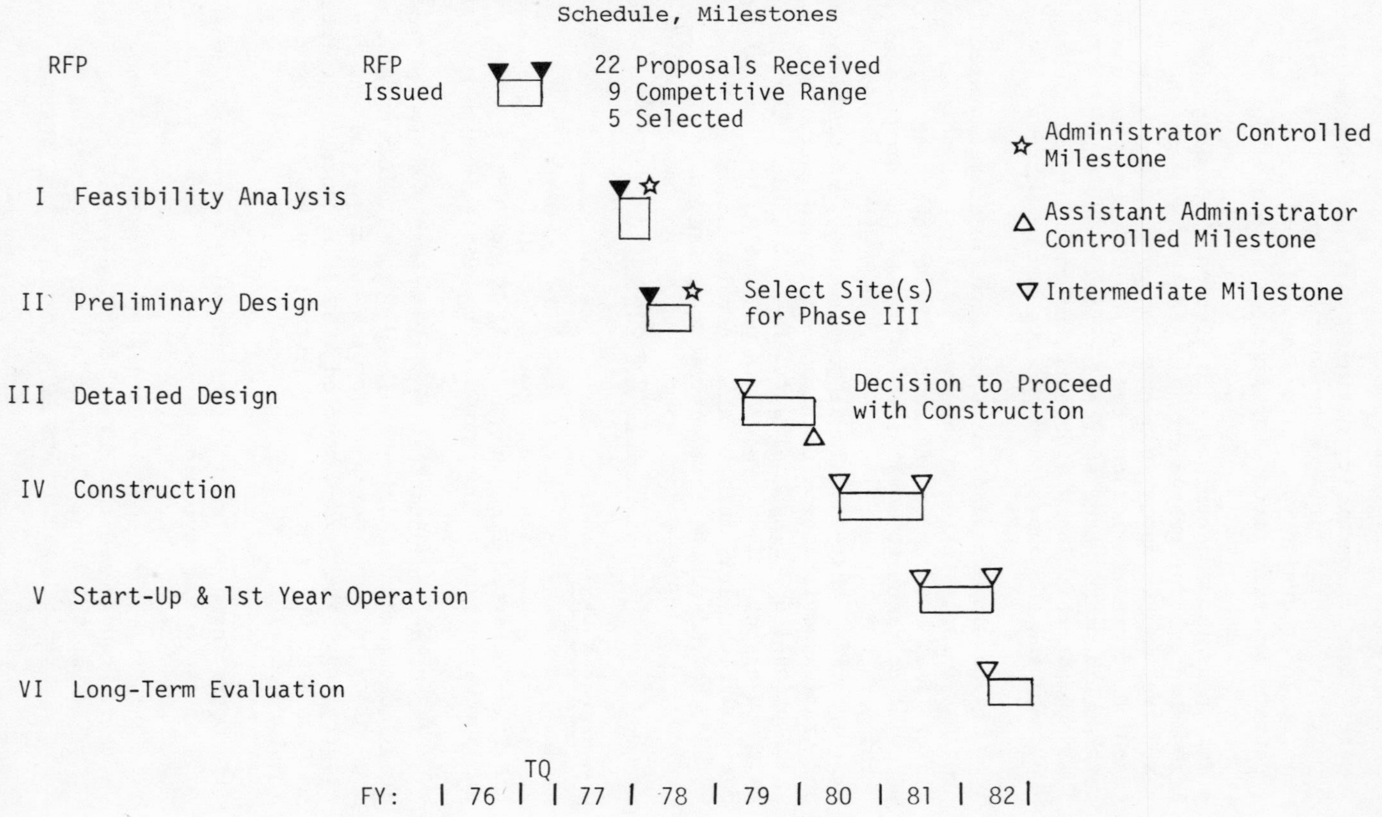

Schedule, Milestones

RFP   RFP
             Issued

             22 Proposals Received
             9 Competitive Range
             5 Selected

I   Feasibility Analysis

II Preliminary Design   Select Site(s)
                                for Phase III

III Detailed Design   Decision to Proceed
                                with Construction

IV Construction

V   Start-Up & 1st Year Operation

VI Long-Term Evaluation

FY: | 76 | TQ | 77 | 78 | 79 | 80 | 81 | 82 |

☆ Administrator Controlled
  Milestone

△ Assistant Administrator
  Controlled Milestone

▽ Intermediate Milestone

electric utility making up deficits or absorbing
surpluses (credit/debit mode).

Because the Grid-Connected ICES operates only in response to thermal demand, there is, by definition, no usable
heat energy wasted at any time. Moreover, the Grid-Connected
ICES is a concept for system operation; i.e., there are many
possible community energy system configurations (types and
layout of equipment for central plant, distribution, and
end-use) available to the designer to develop such a system.
Grid-Connected ICES can be applied today in a wide variety
of communities, including university campuses, medical complexes, downtown renewal projects, new residential or commercial developments, and military bases. They can be used
with existing power companies and should provide annual
energy efficiencies of 70 to 85 percent. The projected
energy savings by 1985 are the equivalent of 9 million
barrels of oil per day.

The grid-connected ICES demonstration program consists
of six phases:

I.   A preliminary feasibility analysis and evaluation of
      candidate demonstration communities.

II.   A detailed feasibility analysis and preliminary design
      for four selected demonstration systems.

III. A final design and preparation of construction working
      drawings for the demonstration system(s).

IV.   Construction of the demonstration system(s).

V.   Initial operation, testing, and performance monitoring
      of the system(s).

VI.   Long-term system operation and evaluation.

The program schedule is shown in Figure 3. Five teams
and sites were examined during Phase I, which lasted from
February 1 to May 30, 1977 (Table 3).

A summary of the five sites examined in Phase I is
presented in Table 4. Projected cost and energy information for each of the sites is presented in Table 5. Four
of these sites and teams (Clark University, Trenton, New
Orleans, and University of Minnesota) have been selected
for the next phase.

The goal of the Advanced Coal-Using Community Systems

Table 3. Five Teams and Sites Examined during Phase I

COMMUNITY	ICES	TEAM
Commercial/Residential 2.1 million square feet	Coal Boilers	City of Independence, NUS, PTC
Small Campus 1 million square feet	Diesel Engines	Clark University, Thermo Electron, F&T Consultants, New England Electric System, State of Massachusetts
Downtown Commercial/Residential 2 million square feet	Gas Turbines	City of Trenton, NCIA, NTC, PSE&G, R. G. Stein, Associates, R. G. Vanderweil Engineers[1]
Hospitals and Health Educational Facilities, 7.5 million sq. feet	Coal Boilers and Steam Turbines	HEAL, NOPSI de Laureal Engineers, OSM & Associates[2]
Large Campus and Adjacent Hospitals 14 million square feet	Coal Boilers and Solid Waste Gasifiers and Steam Turbines	University of Minnesota, N. S. Power, St. Mary's Hospital, Fairview Hospital[3]

[1]The City of Trenton and two State agencies, the Mercer County Improvement Authority (MCIA) and the New Trenton Corporation (NTC), own the demonstration community. The Public Service Electric and Gas Company (PSE&G) will operate ICES.

[2]The Health Education Authority of Louisiana (HEAL) is the owner of the demonstration community. New Orleans Public Service, Incorporated (NOPSI) will operate ICES.

[3]The University of Minnesota Campus in Downtown Minneapolis, with two adjacent hospitals, is the demonstration community. Northern States Power Company (NSP) is the local power company. NSP will sell the power plant to the University which will be utilized as the ICES plant. The University will operate ICES.

Table 4. Grid Connected ICES:
Summary of Phase 1

Community	Type	Size	Proposed G-C ICES
City of Independence Independence, Missouri	Commercial Residential Shopping Center	2.1 Million Sq. Ft. 20 MWe	Coal Boilers
Clark University Worcester Massachusetts	Small Campus with Education and Residential Buildings	1 Million Sq. Ft. 1.5 MWe 20,000 Lbs./Hr. Steam	Diesel Engines (Distillate or Residual)
City of Trenton Trenton, New Jersey	Downtown Redevelopment-Commercial, Institutional, Residential Buildings	2 Million Sq. Ft. 10 MWe 120,000 Lbs. Steam/Hr.	Gas (Turbines or Residual Fueled Systems)
HEAL New Orleans, Louisiana	Hospitals and Health Educational Facilities	3 Million Sq. Ft. Existing 7.5 Million Sq. Ft. Ultimate 11.7 MWe 157,000 Lbs./Hr. Steam 12,000 T Cooling	Coal Boilers
University of Minnesota Minneapolis, Minnesota	Large Campus and Adjacent Hospitals	14 Million Sq. Ft. 7.5 to 12.5 MWe 300,000 Lbs./Hr. Steam	Coal Boilers and Solid Waste Pyrolysis

Table 5. Grid Connected Integrated Community Energy System:
Energy and Cost Information

Site	System Electrical Capacity MWe	Energy Savings Barrels of Oil/Day Equivalent	System Efficiency %	Total Project Cost	Total Project Cost
Independence	20	11	38	38	11.0
Clark University	1.5	22	68	2	1.2
Trenton	10	66	63	14	2.6
HEAL	11	342	78	31	2.5
University of Minn.*	12.5	156-315	55-78	17-25	2.3-5.6

*Lower Figure for Basic Configuration. Higher Savings with Expanded Service Area; Pyrolysis
of Urban Waste; Conversion from Steam to Hot Water Distribution.

Program is to develop community-level, coal-using systems thereby replacing oil and gas as primary fuels for the satisfaction of community energy needs. Principal benefits include: maximum utilization of an abundant fuel (i.e., coal), minimum waste of energy, minimum adverse environmental impact, and economical utility rates to customers.

Because the private sector is not willing to undertake the difficult and time-consuming design and analysis effort, DOE's assumption of the responsibility and the subsequent system demonstration will serve as major incentives for widespread utilization of coal-using systems. Widespread usage will also result in beneficial effects on the national energy balance.

The Coal-Using ICES concept is being applied at Georgetown University in downtown Washington, DC. Phase I of this effort, now underway, is directed at the development of design methods for determining optimal ICES using coal and coal-derived fuels. Coal-using ICES have the potential for co-firing with solid and liquid waste and will use advanced conversion technologies such as fluidized bed in the Georgetown demonstration. The specific tasks being conducted in this Phase I effort are:

1. Evaluate and characterize technologies needed in coal-based community systems (central plant conversion, secondary conversion, storage, transmission and distribution, end-use conversion).

2. Collect and evaluate load data for reference and test communities.

3. Develop system design methods of synthesis and optimization.

4. Develop user-oriented version of computer program for synthesis and optimization of coal-using ICES.

Future phases include developing site-specific design for system(s), construction of specific system(s), and operating and demonstration specific system(s). Other community targets and energy system alternatives are being considered. These potential ICES markets and applications are listed in Table 6.

Community Design

Community planning and design program activities at DOE address the energy benefits to be derived by applying

Table 6. Markets for ICES Applications

COMMUNITY TARGETS

- Federal Facility
  - Government Buildings
  - Military Bases

- Federally Funded Community Development Programs
  - Hospitals/Retirement/Nursing
  - Revenue Sharing
  - University/Colleges
  - Housing Programs
  - Urban Renewal
  - Transportation Facility

- State/Local Government Complexes

- Regional High Density Metrocenters

- New Communities, Urban Development, Other Major Development (e.g. Retirement, Recreation)

- Shopping Centers Commercial Complexes Industrial Parks

- Medium or High Density Residential or Planned Unit Developments

- Agricultural Centers

ENERGY SYSTEMS ALTERNATIVES

- District Heating Systems

- Utility Heating/Cooling/ Cogeneration Systems

- Utility Grid-Connected Integrated Community Energy Systems

- Upgrade Total Energy Applications to ICES

- Integrated Energy, Solid Waste and Wastewater Treatment Systems

- Heat Pump Centered ICES

- Coal Using ICES

- Advanced Technology ICES

---

community design principles and concepts to ICES applications. One of the major program elements in this area is energy conserving General Development and ICES Master Plans. Research efforts are aimed at facilitating, through successful demonstration, the incorporation of energy conservation goals, objectives, policies, programs and technology in the community development and ICES master plans. The primary objectives of the program are:

- To develop, test and demonstrate the integration of energy conservation systems, methods, techniques and mechanisms.

- To initiate policies and actions to overcome existing institutional barriers for the utilization and promotion of energy conservation alternatives.

- Provide laboratories to test energy conservation measures that could be easily transferred and disseminated to other communities.

This research is being carried out in case studies in representative cities.

A case study in Mercer County, North Dakota, will serve as the first rural Community Energy Conservation Laboratory.

We have assisted the elected officials of Mercer County in drafting a joint powers agreement for the establishment of an energy development board. The board is comprised of 6 cities, 5 school districts, and the county. The board will coordinate planning and energy conservation demonstrations in the county, and provide an ideal mechanism for facilitating the rural energy laboratory. Specifically, activities will focus on establishing institutional mechanisms for coordination, development of comprehensive plan for development, and identification of energy conservation opportunities for Mercer County. The Mercer County development management mechanism Energy Development Board could be applied to urban areas as well.

It is our intention to increase the number of case studies, providing a greater diversity of demonstrations, from which many cities can profit in undertaking energy conservation as part of their own master planning activities.

Implementation Mechanisms

The implementation mechanisms aspects of the DOE program are designed to prepare public and private sectors to implement ICES concepts and program products. The rate

Table 7. Factors Influencing the Commercialization of Community Energy Systems

I.   FINANCIAL

      1. Cost

      2. Financing Methods

II.   PUBLIC INSTITUTIONS

      1. Local Planning

      2. Administration

III. REGULATORY

IV.   PRIVATE INSTITUTIONS

      1. Appraisal Practices

      2. Building Costs

      3. Labor Unions

V.   CONSUMER RESPONSE

VI.   SUPPLY

      1. Land

      2. Materials/Equipment

VII. MANAGEMENT/OPERATIONS

Table 8. Community Systems Program

Community Systems Program: Major Elements	Community			Time Frame for Impact		
	Site	City	Metro Region	Near	Mid	Long
Grid-Connected ICES	X			X		
Coal-Using ICES	X	X			X	
Heat Pump Centered ICES	X			X		
District Heating & Cooling		X	X		X	
Total Energy Plant Retrofit to ICES	X	X	X	X		
Advanced Technology ICES	X	X	X		X	X
Systems Engineering for ICES	X	X	X	X	X	X
Site & Neighborhood Design	X	X		X		
Community Subsystems Development Process			X	X		
General Development & Master Plans		X	X			X
Analytical Tools for Community Design		X	X			
Impacts of Regulation	X	X	X	X		
Innovative Mechanisms for Financing and Managing ICES	X	X		X		
Market Analysis for ICES Targets of Opportunity	X			X		
Community Capacity Building for Energy Management	X	X	X	X		
Comprehensive Energy Planning Demonstration		X			X	

of commercial acceptance and deployment of energy-efficient community designs and energy supply technology depends not only on technical feasibility and cost, but also on a number of institutional factors and practices that can affect their marketability. The major factors influencing the commercialization of community energy systems are categorized in Table 7. Among the institutional impediments to the adoption of ICES is the impact of public utility regulations. The determination of their impact is a major program element of the ICES Program.

Integrated systems in most instances involve services that are classified as public utilities and, as such, are subject to some form of public control. Therefore the normal competitive conditions (supply/demand, unrestricted market entry, etc.) do not generally exist. The existing structure of public regulation is geared to individually operated systems such as gas, electricity, telecommunication, and transportation. Planned, integrated community energy systems will be operated as a unified system and may, therefore, be covered by more than one public regulatory body whose rules may require modification to permit the most energy- and cost-effective operation. Program activities are aimed at developing proposals for adoption by the necessary public regulatory bodies to facilitate the implementation of integrated systems. Current efforts are studying and analyzing the impact of existing regulations. Proposed changes in regulation at the Federal, State and local level will be developed. Future efforts will be directed toward the application of the proposed changes to the regulation of integrated systems of various sizes, technologies, and governmental and institutional arrangements.

Current Integrated Community Energy Systems Programs being conducted by the Department of Energy are listed in Table 8. For each program, the community size of potential application and the time frame of impact are presented.

Conclusion

Integrated Community Energy Systems hold great promise as a means to conserving energy and natural resources while providing the energy needed for industry and community services. The Department of Energy is taking a lead role in developing viable Integrated Community Energy Systems, which will be implemented in the private sector at the earliest possible time.

2

Simulation of the Community Annual Storage Energy System

W. Richard Powell

Abstract

The Community Annual Storage Energy System, CASES, is a thermal utility that provides efficient and economical heating and cooling services to a community. Heat is collected in summer and returned in winter to reduce fuel consumption. Winter's cold is retained and used to eliminate conventional air conditioning. The cost and performance characteristic of CASES have been evaluated via hourly simulation studies. As much as 80% of the energy required for heating and cooling of buildings can be obtained "energy free" with lower cost heating and cooling than is common today.

Introduction

Traditionally, we regard heating and cooling of buildings as unrelated problems. In summer we consume electric power to run air conditioners that pump heat out of buildings and throw it away. In winter we usually consume a precious and irreplaceable fossil fuel to generate about the same amount of heat as we threw away the previous summer. Figure 1 shows the typical seasonal pattern of heating and cooling. This is a wasteful cycle, left over from the era of cheap energy, and we can no longer afford it. We must learn to recycle our summer heat instead of throwing it away. This is the concept presented here.

The Community Annual Storage Energy System, CASES, is a new form of solar heating. It uses the buildings themselves to collect solar heat. This gives CASES a tremendous economic advantage over conventional solar energy systems, which uses expensive solar collectors. CASES uses the surplus heat

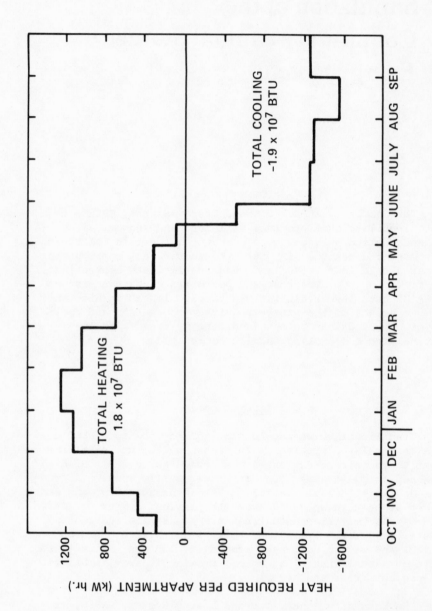

Figure 1. Seasonal record of heating and cooling in a Washington, D.C. apartment complex

removed from buildings in summer as a primary source of
heat for winter. CASES also collects and uses the excess
heat produced in some community buildings even in winter to
further reduce fuel consumption. At times, CASES obtains
a portion of the heat required by the community directly
from the winter environment. By substituting stored summer
heat, excess heat, and environmental heat for conventional
heating fuels, the consumption of scarce resources can be
greatly reduced.

The heat collection fluid used in summer to remove excess heat from a building, must enter and leave the building
at a temperature less than the temperature desired in the
building. Likewise, when this heat is returned to the buildings in winter, the heat transfer fluid used must be hotter
than the building. Thus, a heat pump is required to elevate
the temperature of the heat transfer fluid. CASES uses
water as the heat transfer fluid and storage medium. CASES
also uses water-source heat pumps to elevate the temperature of the heat transfer fluid.

CASES is fundamentally an annual storage energy system.
Thermal energy storage is more economical and efficient on
a community scale than in individual buildings because the
surface-to-volume ratio is more favorable in large scale.
Because CASES has large storage facilities available, it can
draw upon stored energy at times of peak electric power
demand. It has the potential for significantly improving
both diurnal and seasonal load curves of electric utilities.
Unlike conventional solar energy systems, CASES can reduce
the need for standby generation facilities and lower the
cost of electric energy.

Description of System

CASES is a thermal utility plan which provides heating
and cooling services to the buildings of a community via
water pipes from a central energy storage facility. Two
pipelines are used. One is for cold (5 $\pm$ 4° C) water and
the other is for warm (15 $\pm$ 5° C) water. For heating buildings, water-source heat pumps take water from the warm pipe
and extract heat from it to maintain each building at the
temperature set on its thermostat. The chilled water produced by these heat pumps as heat is extracted is injected
isothermally into the cold-water pipeline.

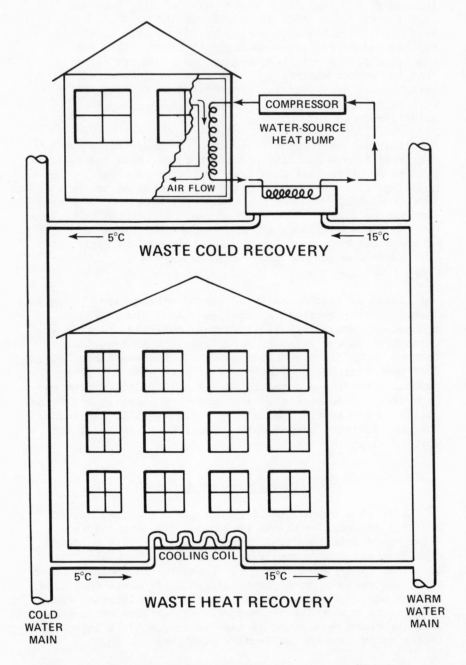

Figure 2.   Synergistic interactions between buildings in a
            CASES community saves energy and cost.

Buildings which require cooling take cold water from the cold pipe and pass it through cooling coils where it is warmed by the excess building heat to the temperature of the warm pipeline which removes it from the building.

Figure 2 illustrates how CASES works using a specific example. Every building in the community has access to both the warm and cold water distribution lines and can draw from them any time it wishes. In fact, a building can draw warm water to heat its north side at the same time it is drawing cold water to cool its south side. Both heating and cooling are continuously available and strictly under the control of the building occupants.

As illustrated in Figure 2, different buildings in the community have diverse heating and cooling requirements. Typically, large buildings with high occupancy require cooling even in winter. Thus, part of the warm water required for heat pumps is supplied by buildings with excess heat. At present, this heat is wasted. Some conventional buildings actually use air conditioners in winter, but typically, cold outside air is taken into buildings with excess internal heat and warm air is discharged.

Even without any use of storage, CASES can save a great deal of the energy which is now wasted. Note in Figure 2 that the cold water produced as a by-product of heating the house is immediately used to provide cooling to a larger building. We call this "waste cold recovery." An ordinary air-source heat pump simply discards the cold air it produces. Waste cold recovery is very closely related to the concept of waste heat recovery, which is illustrated by the warm water emerging from the larger building shown in Figure 2. If this building were cooled by an air conditioner, not only would additional electrical energy be required, but the heat pumped out of the building would be wasted. In CASES this wasted heat is recovered and used to help heat the smaller building more efficiently. Thus, with both waste heat recovery and waste cold recovery, it is possible to heat the small buildings and cool the larger buildings in a community with less electric power than would be required simply for heating alone if ordinary air-source heat pumps were used.

Aquifer Storage

We cannot always immediately utilize the waste heat and waste cold that is available. Typically, in winter there is an excess of waste cold and in summer there is an excess of waste heat. Figures 3 and 4 show one of the more

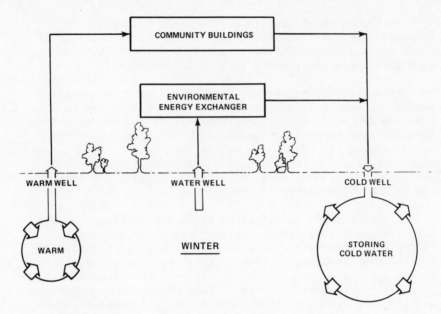

Figure 3.   Aquifer flow pattern in winter

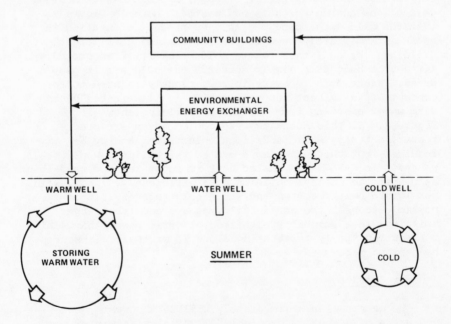

Figure 4.   Aquifer flow pattern in summer

attractive ways to store the excess warm and cold water.
The most economical way to save the large volume of warm
water produced during summer as heat that is removed from
community buildings is to inject it into the ground. The
earth is a very good insulator. If the geological characteristics of the site are suitable, i.e., if a porous water-
storage aquifer exists under the community, then the summer
heat can be stored with little loss until the community
requires warm water for heating. During the winter, this
stored warm water is withdrawn and delivered to the heat
pumps for heating buildings. As heat is removed, cold water
is produced and returned to the central CASES facility via
the cold water pipelines. During the winter this flow of
cold water is injected into the ground and stored in a
second aquifer or a separate region of the same aquifer used
to store warm water. For every pound of warm water withdrawn from the ground, a pound of cold water is concurrently
injected into the ground. Thus CASES does not consume
ground water.

We may collect more cold water in winter using natural
cooling and inject it along with the cold water produced by
the heat pumps in the community as illustrated in Figure 3.
This is to insure that there will be enough cold water for
non-electrical air conditioning even if the current winter
is a mild one and the following summer is unusually hot.
By the end of winter we have a lot of cold water in storage,
but we are starting to get low on warm water. At the end
of summer, the situation is just the reverse. We have drawn
down our store of cold water and built up the supply of warm
water as shown in Figure 4.

CASES uses the ground water as a "heat bank". During
the summer when warm water is readily available, CASES
deposits heat into the ground-water heat bank. During the
winter when heat is needed, CASES withdraws heat from the
ground-water heat bank.

Ice Storage

Where the geology or other factors do not permit the
annual storage component of CASES to be based on aquifers,
it is necessary to provide an artificial method of storing
the surplus summer heat and winter cold until they are
needed. If suitable aquifers are unavailable, it is not
economically feasible to store all the cold water produced
in winter as a byproduct of heating with water-source heat
pumps. Instead, the cold water flowing back to the central
CASES facility is heated and returned to the community in a
closed-loop flow pattern. The heat required can be obtained

from a pond with an ice machine.

A great quantity of heat must be removed from liquid water to produce ice. For each cubic foot of cold water converted into ice by an ice machine, more than 10,000 Btu are made available for reheating the water circulating in the community distribution pipelines. Ice machines taking heat out of a cold water pond will deliver more heat to the circulating water than could be obtained directly from the same pond if it were full of boiling water! Ice machines are simply heat pumps which also produce ice.

The ice produced during winter is saved for non-electrical "energy-free" cooling the following summer. It is stored under a floating insulated roof in the same ice/water pond from which the ice machines take heat in winter. As the winter progresses, more and more of the ice/water pond is converted into ice. During summer, this ice is melted while cooling the community. The ice/water pond is made large enough to store all the ice produced in winter and desired for summer cooling.

System Components

Environmental Energy Exchanger

"Energy-free" heat can be collected directly from the environment with an Environmental Energy Exchanger, EEE. The EEE may resemble a dry cooling tower if it functions primarily as a heat collector. It can be much smaller than the solar collector required to heat the same community because it only collects the annual imbalance in system heating requirements and not all of heat used by community buildings. The EEE can collect heat any time the air temperature is greater than the cold water temperature (~5°C). It can often be used 24 hours a day and is always more efficient than a solar collector because of its low operating temperature.

In northern communities the EEE can collect heat to melt ice in the ice/water pond any time the air temperature is above freezing. By melting excess ice during the winter it is possible to freeze the water in the ice/water pond several times each winter and obtain as much heat as if the pond were several times larger. Thus, in northern locations, the EEE can be used both to substitute "energy-free" environmental heat for ice-machine heat and to reduce the size and cost of the ice/water pond required.

In southern locations, the heat removed from buildings in summer exceeds the heat required in winter. In this case, the EEE might resemble a spray water pond. The EEE can be used to discharge this excess heat in winter without using electrically driven air conditioners if aquifer storage is feasible. See Figure 3. If an ice/water pond is used for annual storage, the EEE can be designed to collect natural ice in winter to supplement the ice produced as a byproduct of winter heating.

Thus the EEE is used to achieve an annual balance in the heating and cooling loads and to reduce both energy consumption and costs. An EEE which can be used for both ice and heat collection is illustrated in Figure 5 as part of the central CASES facility which is schematically represented in Figure 6.

Water Source Heat Pumps

Efficiency. The water-source heat pumps used in CASES buildings remain highly efficient even in cold weather because warm water from the pipeline is always available to them as a source of heat. In cold weather air-source heat pumps tend to condense water vapor from the air and freeze up. Periodic defrost cycles are required. Air-source heat pumps become less efficient as the weather becomes colder. In subfreezing weather, most air-source heat pumps revert to electrical resistance heating and have a coefficient of performance, COP, of unity or less; but the COP of CASES heat pumps typically exceeds five at all times. It should be noted that although some energy is consumed at the central CASES facilities to keep warm water continuously available, much of the warm water is solar heat saved from the summer, or "energy-free" heat collected in winter either by the EEE or while cooling the larger buildings of the community as illustrated in Figure 2.

Costs. The water-source heat pumps used in CASES are similar to air-source heat pumps except their input heat exchanger is smaller and no outside fan-coil unit is required. They should be more economical than air-source heat pumps when produced in the same volume. The absence of outside fan-coil units reduces installation costs and makes areas outside of buildings more useful.

Peak Loads. Air-source heat pumps require less electrical energy than electrical resistance heaters on mild winter days but often require slightly more on cold winter days. They thus represent a more variable electric load

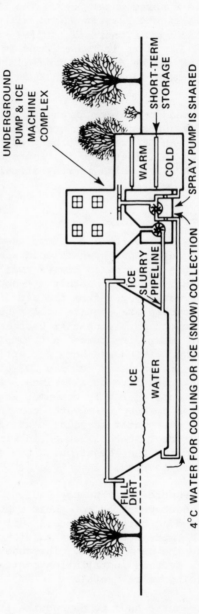

Figure 5. Ice/water pond annual storage system with ice or heat collection via spray-type enviromental energy exchanger

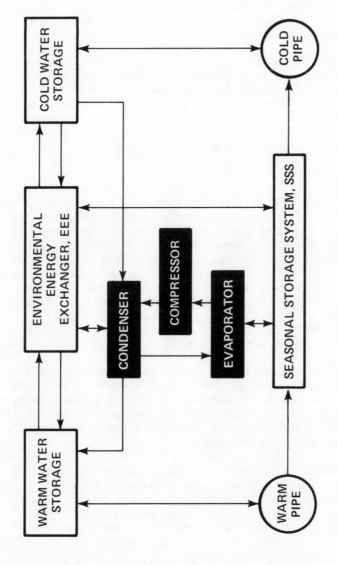

Figure 6. Heat and fluid flow diagram for CASES central plant with ice-machine heat pump

than electric heaters and worsen electric utility load
curves (equipment utilization averages). Thus, like
conventional solar heating, wind-energy systems, and other
"part-time" energy systems with electrical backup, air-
source heat pumps tend to make electric power more expensive.
Someone must pay for idle electrical generation equipment
held in reserve as backup for part-time energy systems.

The water-source heat pumps used in CASES also require
more electric power on cold days to supply more heat but
the peak electric demand is typically five times smaller
than with an air-source heat pump. Thus, peak heating loads
can be reduced 80%. CASES can further improve the average
utilization of electric utility equipment by actions taken
at the central CASES facility as discussed below.

Short-Term Storage

The central CASES facility also provides for short-term
storage of surplus heat collected from the community. The
short-term storage facility consists of two water impound-
ment basins of equal volume. One is used to hold cold water
and the other for warm water. They are used to smooth out
or buffer the diurnal fluctuation in the demand for heating
and/or cooling waters. Typically, their volume is large
enough to permit a major portion of the demand for heating
water during the hours of peak electric usage on cold
winter days to be satisfied directly from stored warm water
with minimal use of electric power.

Often in the spring and fall, the community requires
cold water for cooling in the late afternoon and warm water
for heating in the early morning hours, but relatively little
average flow. In this case, the short-term storage facility
can significantly reduce the energy required to supply the
thermal demands of the community.

Even when aquifers are available, short-term storage
facilities on the surface may be desirable to buffer the
flow of warm and cold water into the aquifers. The well
pumps and well-field bore-hole expense can be reduced if
part of the peak flow is into and out of water storage
basins on the surface. An ice-collecting EEE may be useful
even with aquifer storage if coupled to these short-term
storage facilities. On the coldest winter nights, ice can
be rapidly collected and then converted into cold water for
injection into the aquifer at lower flow rates. Excessive
flow rates can damage the aquifer and result in greater
pumping power requirements.

The savings in peak power or "demand" charges possible with short-term storage facilities are most important when aquifers are not available. It is possible to operate the ice machines at night or other times when electric power is cheap with considerable economic advantage if short-term storage facilities are included in CASES.

Many electric utility companies have recognized that the cost of thermal storage can be less than the cost of expanding peak generation and transmission facilities which would be required in the absence of thermal storage facil- ities located near the customer. With community-scale thermal storage as used in CASES, the cost of these storage faciliites can be reduced because the large scale results in a more economical surface to volume ratio than storage in individual buildings. Also, the cost of separate elec- tric meters for off-peak electric power is reduced with centralized storage. Only one meter for each rate period is required for the entire community. It is economically feasible to provide a separate meter and rate scale for each hour of the day, if desired, to reflect more accurately the electric utility cost of production. Present "time-of- day" prices are limited to a few different rates to avoid excessive costs for the meters installed in each building and only crudely reflect the true cost of electric power as a function of time.

The short-term storage facilities available at the central CASES plant could be operated directly by the local electric utility to produce a more cost-effective form of load control than time-of-day rates. Prearranged electric rate schedules, regardless of how detailed, cannot reflect unforeseen events which effect generation capacity and costs. The ability to take all ice-machine heat pumps "off-the-line" for a few hours in power emergencies could prevent a "black- out" in the community. During this period all of the thermal output from the central plant would be drawn from short-term storage facilities.

Although the short-term storage facilities can effec- tively reduce energy consumption in spring and fall and shift wintertime electrical energy consumption from daytime peak demand periods into the night time valleys of electric utility load curves, they are unable to cope with the pro- longed periods in summer when the community demands only cold water for cooling. Long-term or annual storage is required to save this surplus summer heat until winter when it can be utilized.

Long-Term Storage

As discussed earlier, aquifers usually represent the most economical and efficient form of long-term storage, but the ice/water pond alternative with its associated ice machines provides a greater opportunity to improve the average utilization of electric utility equipment. Electric utility companies usually have a seasonal imbalance in the demand for electric power as well as diurnal peaks in the electric demand curves. For a utility company with a grow-ing summer peak load, the fact that CASES provides air conditioning non-electrically either by cold water from an aquifer or by melting ice stored from winter can reduce the chance of summer blackouts. The fact that CASES may replace present gas and oil furnaces with electrically powered heat pumps is also attractive as this adds electric demand in winter. This shift of electric demand from summer to winter is a direct result of the annual storage feature inherent in CASES. It can lower the price of electricity in a summer peaking electric grid.

For an electric utility with a wintertime peak demand, the surplus heat collected in summer is the attractive long-term storage feature. Together with proper use of short-term storage facilities for diurnal load management, it permits the peak electric heating demand to be about five times less than if air-source heat pumps were used. The lower peak demand and higher base load possible with CASES can lower the cost of electric power in a winter peaking electric grid.

Water Distribution Pipelines

Although CASES needs water distribution lines, they require little energy and add only a small cost to the system in a new community. Because the warm and cold water circulating in these lines is only a few degrees different from ground temperatures, it does not appear cost effective to insulate these pipelines. They would be essentially the same in construction as potable water lines.

The heat lost (or gained) during distribution has not been evaluated in detail yet because accurate analysis is complex. Preliminary estimates indicate that it will be less than 10% of the heat delivered. It should be noted that most of this low temperature heat (and cold) is "energy-free". These thermal losses have little economic significance, but may require that slightly more summer heat and winter cold be saved.

The cost of thermal water distribution should add less
to the cost of heating and cooling than current water bill
because the water used in CASES pipelines is not consumed
nor is it processed to potable water standards. In many
communities the cost of the sewer lines and sewage treatment
plants is also included in the water bill. For these communities the cost associated with distribution pipelines for
CASES should be much less than their current "water" bills.

Ice Machines

The ice machines used in CASES produce flake ice
because heat transport through thick ice would make them
less efficient and more costly heat pumps. Flake ice is
also convenient since it can be pumped in an ice/water
slurry to the bottom of the ice/water storage pond. These
ice machines are not needed for heating in summer and can
be used to manufacture ice for sale to defray part of their
costs with a corresponding reduction in the cost of heating
and cooling services.

Analysis

Method

Proper evaluation of any annual storage energy system
requires data for a full year. The cost and performance
characteristics of CASES have been determined using a
computer simulation to model 8760 hours (one year) of se-
quential steady states. The simulation model assumes that
an ice/water pond is used for annual storage instead of
aquifers. It thus sets an upper limit on costs and a lower
limit on efficiency since ice machines are both more expensive and less efficient than pumping warm and cold water
into the ground. In other ways, the model is conservative
in its estimates and assumptions so that future refinements
which make it more complete and detailed should not lead to
any significantly less attractive results.

CASES Model

The first of four computer program modules contained
in the CASES simulation is called HCLOAD. This program
produces a file of hourly heating and cooling loads (Btu/h)
for each of nine building types and two core zones. In
addition to the hourly output data concerning heating and
cooling requirements, certain peak and total loads are
recorded. These auxiliary data are used in subsequent program modules to determine efficiently both the size of heating and cooling equipment required in each building and the

size of various energy storage facilities needed at the central CASES plant.

The second module of the simulation code, CDIST, requires the distribution water temperatures in addition to the files written by HCLOAD as input data. CDIST calculates the size and cost of user heating and cooling equipment needed for each building. It produces an hourly record of the electric power required for operating this heating and cooling equipment. It also converts the thermal load data into a demand (pounds of water per hour) data file. This conversion depends upon the water temperatures and the efficiency of the equipment selected for each building.

CDIST uses the water-demand file and data on the pipeline route to each terminal building to determine the size and capital cost of all pipes and trenches in the distribution system. More refined editions of CDIST also calculate pumping and thermal losses and the associated costs. The current simulation model doubles costs of the pipe and trench given in Table 1 to form a conservative estimate of pipeline and other cost details omitted from the distribution system. Finally, CDIST reduces all the demand data down to a single net-demand file.

Table 1
Cost of Pipes and Trench
(uninsulated, asbestos-cement
pipes in a new community)

Pipe Diameter (Inches)*	Installed Cost ($/Foot)*
3	$ 10.94
4	$ 12.03
6	$ 13.68
8	$ 16.10
10	$ 19.71
12	$ 22.40
14	$ 27.20
16	$ 32.54
18	$ 39.12
24	$ 63.82
30	$ 88.52
36	$113.76

* SIU not convenient units for this commercial data

The third simulation module, CAPS, simulates hourly operation of the central CASES plant as it services this net demand under an assumed set of operating rules. CAPS determines the size and cost of the various energy storage facilities and ice machines required at the central plant. CAPS also computes the hourly requirements for electrical power.

The final module, COST (still in progress), will process all the various cost and energy consumption files written by CDIST and CAPS. Various financial rates and Btu prices will be used in COST along with the processed CASES data to determine the total cost of CASES and the expected rate of return on invested capital. COST will also total all energy consumption and compare it to the heating and cooling services supplied in order to evaluate the efficiency and annual energy savings. Presently, hand calculations have been used to combine the results of CDIST and CAPS. In lieu of detailed economic analysis, it is assumed that 15% of the capital is recovered annually.

Reference Community

It is necessary to specify the community in considerable detail to evaluate energy costs. The selected model community is similar to Wilde Lake Village in Columbia, MD. It has 8000 residents (2500 households) and 866 buildings in three neighborhoods that share a common village center, all on 2500 acres of land. Each neighborhood has its own elementary school and civic-commercial center. The community's middle school, high school, office buildings, factory, shopping center, and CASES facility are located in the village center. These buildings and clusters of residential units comprise the 43 zones (local load centers) shown in Figure 7. Figure 7 also illustrates the route of major distribution lines, but not the connections to individual buildings in each zone. The composition and demography of the reference community are further described in Tables 2 and 3. Hourly weather records for 1967 were used.

Results

Predictions of efficiency and costs for the standard reference community are given in Tables 4 and 5. Forty-three percent of all the cooling water required during the year was waste cold recovered from cooling other buildings as illustrated in Figure 2. Twenty-eight percent of all warm water required was energy-free waste heat. However, because heating is dominate, the total waste heat recovery exceeds the total waste cold recovery. The overall coeffi-

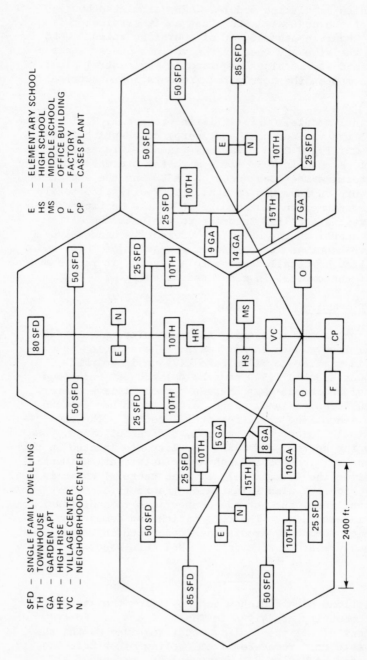

SFD — SINGLE FAMILY DWELLING
TH — TOWNHOUSE
GA — GARDEN APT
HR — HIGH RISE
VC — VILLAGE CENTER
N — NEIGHBORHOOD CENTER

E — ELEMENTARY SCHOOL
HS — HIGH SCHOOL
MS — MIDDLE SCHOOL
O — OFFICE BUILDING
F — FACTORY
CP — CASES PLANT

2400 ft.

Figure 7. Layout of 866 buildings and pipeline route in standard reference community

Table 2
Community Buildings

400	Two-story, Four-bedroom Frame Houses
300	One-story, Three-bedroom Brick Houses
100	Two-story, Four-unit Townhouses
53	Three-story, 24-unit Apartments
1	14-story, 140-unit Apartment Building
5	One-story Schools (different sizes)
2	12-story, 1000-worker Office Buildings
4	One-story, Civic-commercial Buildings
1	One-story, Two-shift, 40,000 sq. ft. Factory

Table 3
Community Demography

2500	Residential Households
4625	Adult Residents
2890	Full-time Jobs and Workers
330	Part-time Jobs and Workers
933	Preschool Children
1320	Primary School Students
561	Middle School Students
561	High School Students
8000	Total Resident Population

cient of performance for the year is 5.5 with a monthly low of 4 and a monthly high of 17.

For this community of 8000 people and 866 buildings it takes about a $15 million investment to build CASES. If a winter peak electric utility company could produce an increment of new capacity at $700/KW$_e$, then it would take $20 million dollars of plant expansion to supply the increase in peak demands that this community represents if it is electrically heated and cooled. Thus it appears that not only can CASES save a great deal of energy but it may be able to conserve investment capital as well.

Based on an analysis of current relative costs, including capital charges and equipment life times, we assumed that charges to remove a Btu from a building should

be about four times higher than to supply a Btu to a
building. With this assumption, the price for heat supplied
is $2.69/MBtu. Thus, CASES can save both fuel and capital
and remain competitive.

Table 4
Monthly Summary for Standard Community

| (1967) | Services Provided | | Efficiency |
	Heating (MBtu)	Cooling (MBtu)	Combined COP
Jan	59,460	15,150	4.509
Feb	69,430	13,130	4.001
Mar	43,240	16,190	5.279
Apr	25,780	17,410	6.707
May	38,170	16,100	5.414
June	8,285	21,110	12.142
July	4,809	22,220	17.216
Aug	7,000	20,210	13.411
Sept	11,120	19,500	10.515
Oct	30,230	17,310	5.783
Nov	49,760	14,880	4.972
Dec	60,580	14,870	4.586
Annual	407,864	208,290	5.524

Table 5
Financial Summary for
Standard Community

Capital Cost at Central Plant	$ 5,097,449
Capital Cost for Pipelines	$ 1,435,000
Capital Cost in Buildings	$ 8,758,400
Total Capital Invested	$15,291,000

(Electrical utility capital excluded)

Total Electric Cost	$ 1,047,000
15% of Capital Total as Annual Cost	$ 2,294,000
Total Revenue Required Annually	$ 3,341,000

In the fiscal analysis of the model we have assumed that the CASES utility owns all of the heat pumps and fan-coil units installed in the community buildings. This building equipment is a major portion of the total capital required for CASES (See Table 5). The cost of heating and cooling services would be significantly less if conventional financing by a building owner was assumed for building equipment. The price of a home heated and cooled by CASES with utility ownership of building equipment would be at least $2,000 less than one for which the owner had to buy and install an air conditioner, a furnace, and a flue-pipe system. Thus, CASES can save fuel, save capital, remain competitive, and lower the purchase price of buildings.

A comparison with conventional solar systems is useful. The capital invested with CASES to provide heating and cooling services to both residential and non-residential buildings in the standard reference community is less than $6,200 per resident household. Typically, a conventional residential solar-electric hybrid system costing this amount would require more electric energy for "back-up" heating on cloudy days and would not provide any cooling. Thus CASES provides both heating and cooling services to all the buildings of a community with less capital and with lower energy consumption than a conventional "heating-only" solar system would provide to only the residential buildings in the community.

Neither the standard reference community nor the design parameters and control strategy assumed have been selected to optimize the cost and performance characteristics of CASES. Work in this area has only recently begun. Table 6 provides some preliminary results which indicate that other communities and other designs for CASES may be more efficient and economical. In the standard reference community it was assumed in the model that the temperature of cold water, TCW, was $4.4^{\circ}C$ ($40^{\circ}F$) but, in view of Table 6, higher cold water temperatures may be more attractive. (The temperature of both the cold and warm water pipelines should vary with the weather for optimum results, but only portions of the computer code required to simulate this variation are currently functional.)

The highest air temperature, HAT, used in the building during peak winter heating demand is $48.9^{\circ}C$ ($120^{\circ}F$) in the standard simulation. A larger flow of less warm air can also deliver the same heat and permit the water-source heat pumps to operate more efficiently as illustrated in Table 6 with HAT = $43.3^{\circ}C$ ($110^{\circ}F$). But the economy of HAT = $43.3^{\circ}C$, predicted by the simulation model, is suspect as the cost increment for larger ducts and/or air fans is not included in the current model.

Table 6
Summary of Results for
Different Communities

Cases Conditions	Annual COP	Heat Price ($/MBtu)	Capital Required ($ per Capita)
Reference Case	5.24	2.69	1,911
TCW = 7.2°C	5.83	2.64	1,908
HAT = 43.3°C	5.58	2.56	1,784
+2 Office Bldg.	5.99	2.44	2,040
+4°F Warmer	6.23	2.59	1,994
Double Size	5.24	2.55	1,759
Double Density	5.24	2.48	1,708

Increasing the total community cooling load relative to the larger total heating load either by adding office buildings to the community or by assuming warmer weather produces a community more nearly in thermal balance on an annual basis. Such a community is more efficient and provides more economical services as indicated in Table 6. The one case studied, two extra office buildings were added to the standard reference community. In the other case, it was assumed that the air temperature was 2.2°C (4°F) warmer each hour of the year.

The effect of doubling both the population and the number of buildings served is also included in Table 6. In one case the density is held constant and the size is increased. In the other case the area of the community is held constant and the population and building density are increased. In both cases there is little change in system efficiency. As one might expect, the larger, denser community is more economical to heat and cool, but the primary reason for the economy of CASES heating and cooling is the large scale, not the high density. The standard reference community is economical with residential density of only one household per half-acre, which is not "high-density" housing. The economy of CASES heating and cooling arises from large scale-cooperation with nature and the synergistic interaction between diverse buildings in the community. CASES is a heating and cooling system hundreds of times larger than that used in an individual residence and it enjoys many benefits of scale even in a low-density community.

Summary

A summary of CASES features is provided in Tables 7, 8, and 9. Table 7 lists items thought to be of interest to the electric utility industry and should be self explanatory in view of preceeding text.

Table 7
Electric Utility Features of CASES

1. Lower Summer Peak Loads
2. Lower Winter Peak Loads
3. Higher Daily Load Factors
4. Higher Electric Consumption (displaces gas and oil heating)
5. Cheaper Electric Power

Table 8
Home Owner Features of CASES

1. Lower Cost Housing
2. Lower Operating Costs
3. No External Units
4. Maintained Automatically
5. Higher Quality Service

Table 8 presents CASES features of primary interest to the home owner (or apartment dweller). The most attractive feature from this point-of-view is the fact that CASES can lower the purchase price of a home compared to a conventional home which requires a furnace, air conditioner, and flue system. With CASES structured as a thermal utility, the home owner rents heating and cooling equipment more efficient and economical to operate than he can provide by acting alone.

Home owners often do not provide any maintenance for their own air conditioners until they become so inefficient that they cannot keep the house cool even if with continuous operation. This irrational behavior of the home owner is neither economical nor efficient, only typical. CASES owned equipment would be efficiently maintained at economically rational intervals and inspected during regularly scheduled visits of the meter reader.

CASE would offer higher quality heating and cooling service than is typically available today with fossil furnaces. It is not safe to operate fossil furnaces at half of their rated output. Consequently, they operate at full output for half of the time. Thus, intermittent heat is delivered at a temperature higher than necessary. With CASES heat pumps of the preferred type, the heat would be continuously supplied at a lower temperature. It takes less energy and provides a higher quality service if gentle, continuous heating is used and this mode has been assumed in the CASES analysis.

Table 9
National Features of CASES

---

1. Lowers Oil Imports
2. Lowers Air Pollution
3. Lowers Capital Requirements
4. Creates Low-skill Jobs
5. Reduces Energy Consumption

Table 9 lists some features of the CASES concept which have national significance. In addition to these benefits, the fact that CASES can be implemented with currently existing technology is significant. All of the components of CASES are commercially available, only the system concept is new.

Acknowledgments

CASES work has been sponsored by the Department of Energy under contract No. 31-109-38-3995 and by the United States Navy under contract N00017-72-C-4401. Neither agency has reviewed this text.

The author wishes to thank his co-workers on this project, S. E. Ciarrocca, D. L. Thayer and G. E. Williams. Without their personal interest and dedicated efforts, it would have been impossible to construct the CASES simulation model.

COMMENTS

Larry S. James (Washington D.C.): Have any comparisons been made between a run-around heat pump system and CASES ? What is the life cycle cost-benefit comparison with a 10% discount rate and a real cost growth of electric power of 2%/year ? Is the storage cost of $35/gal really justifiable ?

Author's Reply: Life-cycle-cost analysis has not been used in CASES studies. We have compared CASES only to other popular electrical systems to avoid speculating on the relative cost of various energy forms. CASES is more efficient than air-source heat pumps and uses a greater fraction of its total electrical input at times when electric power is relatively inexpensive. CASES thus has a lower energy cost. CASES also requires less capital investment than conventional electrical systems when the cost of the electrical generation and distribution system is greater than $600/KW. This is true for new communities regardless of the time of peak electrical demand. Life-cycle-cost analysis is not required to recognize that a less expensive and more efficient system is superior.

Because we do not use life-cycle-cost methods, we have not had to postulate any particular discount rate. To estimate heating and cooling costs, we convert the total capital required to construct CASES and provide the heating and cooling equipment in all community buildings into an equivalent annual cost by assuming 15% of this capital must be collected from the customer in year of analysis. This is a simpler, but more rigorous financial burden for CASES than discounting the stream of future income by 10% each year in a life-cycle-cost analysis.

The cost of thermal water storage facilities increases with size and relatively little data is available in the large volume range contemplated for CASES. The cost data used in the CASES model comes primarily from the 4,000,000 gallon thermal storage basin constructed by Stanford University in 1977. It cost $68,000 but approximately 35% of this cost was for the strong roof and columns in the tank used to hold up the computer building located above the tank. The architect-engineer for the project estimated that it could be one third larger for only a 10% increase in cost. Our analytical cost versus volume function goes through the adjusted Stanford data point with this slope. In this volume range, thermal storage costs about 10¢/gallon. For reservoirs with less than 200,000 gallons, the cost data used in CASES analysis is taken from Table 6.1 of "The Effects of

Thermal Energy Storage on MIUS Cost and Fuel Consumption"
(ORNL/HUD/MIUS-26) with slight modifications to reflect
different lower temperature insulation requirements. With
this data, the 200,000 gallon reservoir costs 35.4¢/gallon,
a 50,000 gallon reservoir costs 83.4¢/gallon, and a 25,000
gallon tank costs $1.39/gallon. To make a smooth connection
between the Stanford and MIUS data, we assume that any reservoir in the 200,000 to 1,000,000 gallon range has a constant 35¢/gallon cost.

If aquifer storage is possible, the effective storage
cost can be only 1¢/gallon. We have tried throughout the
CASES model to use only hard data and commercially available
products even if speculative costs such as aquifer storage
would result in much more favorable results. The economy
inherent in the CASES concept is so strong that attractive
heating and cooling prices result even with unfavorable data
selection and conservative assumptions introduced into the
model for computational simplicity.

The answer to your first question is No. Almost all
of the effort to date has centered on the construction of
the simulation model. Quantative cost and performance data
is now available for comparison with other systems. One
initial comparison study is completed. It is the Task E-1
report "Cost Comparison of CASES with an Air-Source Heat
Pump System". This report was submitted to the sponsor
(DoE/ANL) in April of 1978 and should be approved for public
release by the time this book appears. In essence, the
results of this report show that CASES is more efficient
and requires less total capital investment if all of the
system capital, including electric utility capital, is
taken into account. It should be noted that many assumptions
very favorable to the air-source heat pump were made in this
comparison. For example, it was assumed that the air-source
heat pump had a COP of 2.49 during every hour of the
coldest month and higher COP's at other times. Inspite of
many assumptions biased against CASES, the results of this
comparison were that CASES is both more efficient and costs
less. Cooperating with nature and other buildings via CASES
to reduce energy consumption is an economically sound idea.

3

The Application of District Heating Systems to U.S. Urban Areas

John Karkheck and James Powell

Abstract

In the last few decades district heating systems have been widely used in a number of European countries using waste heat from electric generation or refuse incineration, as well as energy from primary sources such as geothermal wells or fossil fired boilers. The current world status of district heat utilization is summarized. Cost and implementation projections for district heating systems in the U.S. are discussed in comparison with existing modes of space conditioning and domestic water heating. A substantial fraction, i.e., up to ∿ one-half of the U.S. population could employ district heating systems using waste heat, with present population distribution patterns. U.S. energy usage would be reduced by an equivalent of ∿30% of current oil imports. Detailed analyses of a number of urban areas are used to formulate conceptual district energy supply systems, potential implementation levels, and projected energy costs. Important national ancillary economic and social benefits are described; and potential difficulties relating to the implementation of district heating systems in the U.S. are discussed. District heating systems appear very attractive for meeting future U.S. energy needs. The technology is well established. The cost/benefit yield is favorable, and the conservation potential is significant. District heating can be applied in urban and densely populated suburban areas. The remaining demand, in rural and low population density communities, appears to be better suited to other forms of system substitution.

Introduction

The technology of distributing heat from central heat generating plants to a variety of customers for space conditioning, water heating, and industrial applications is quite old. The practice in the United States has been, from the beginning, one-hundred years ago, the generation and distribution of steam to customers in central business districts and some residential areas. For many reasons the district heating industry has not become a major energy supplier in the United States, and in fact appears to be in a decline [1]. The current service level is estimated to be equivalent to 1 percent of the population [2].

In contrast, district heating is flourishing in most European and some Asian countries. There are two principal reasons for this. Appreciation for the energy conservation potential and enhanced environmental quality afforded by central station boilers and by the utilization of power plant and other industrial reject heat has become manifested in policies which encourage the development and use of the technology. Hot water is almost universally used as the heat transport medium. This detail is important both because it is easier to transport water than steam over long distances, hence, water-based large district heating systems are made technically feasible, and for utilization of heat and electricity coproduction since steam extraction is more costly to power plant efficiency than is alteration of the condensing cycle to produce hot water. These features have acted synergistically to provide the impetus for growth of district heating.

Technical Aspects of European Systems

In Western Europe district heating generally includes service to the space and water heat market, whereas, in Eastern Europe process loads may also be served. Various methods are employed to generate useable heat including central station fossil fired boilers, refuse incineration, industrial processes, and power plant turbine cycle modifications. The first source may be used initially, but as systems grow in size, encompassing whole cities in many cases, the demand levels become sufficiently great to utilize the reject heat of power plants specially designed to meet varying electrical and heat loads. Reserve and redundant capacities may then be realized in the former sources. In Iceland and some other countries geo-hydrothermal wells serve as the primary source of hot water for district heating.

Pipe systems are made of welded steel and usually comrise a closed system. Rural transmission lines may be installed above or below ground. Distribution mains in urban areas are always buried, in styles ranging from simple sandfill around insulated pipes to elaborate suspension procedures in precast concrete conduit. This last is becoming increasingly common as a means to keep external water from the pipe to avoid corrosion. With just several inches of standard insulation materials, heat losses may be held to a few percent of the transmitted heat and even this is compensated somewhat by pumping energy. Research and development efforts in many countries are directed toward producing better pipe and insulating materials to reduce maintenance and operating costs and toward reducing installation costs to permit extension of service to low density markets. Sweden has taken the lead toward development of noncorrosive pipe materials, including plastics and polymer concrete.

Building service equipment shows greater variety among nations and even among cities in a given country, due in part to variants in operating conditions and in part to historical conditions. Generally, systems supplied by water at less than 100°C do not require heat exchange equipment for building service, whereas superheated water cannot be fed into typical building equipment. In addition to regulating the temperature and flow rate of users' heat, heat exchangers ensure the closure of the district heating system thus minimize makeup water requirements and protect users' equipment from pressure fluctuations in distribution mains [3]. Metering is a more controversial subject. Heat consumption is difficult to meter because flow rates and inlet and outlet temperatures vary temporally with outdoor temperature and hot water usage. It is expensive to monitor these and experience is showing that such precise information is neither realistic nor necessary. Internal heat flows between apartment or office units belie the measured heat requirement for any unit [4], and for a given building the load is sufficiently well defined to be determined by volume of water used [5]. In the USSR, and probably all of Eastern Europe, floor space and number of occupants are the determinants for billing [6]. Thus, average pricing appears to be the rule.

Virtually all nations of Central, Northern and Eastern Europe, and several in Asia have district heating. For all but Japan, district heating is a municipal enterprise, thus heat supply is treated as a municipal service such as street management, water supply, and sewage disposal. In Eastern Europe, production and distribution are combined enterprises but in Western Europe, power plants are typically owned by a separate authority or company which then sells heat to the district heating authority. In Eastern Europe, district heating and coproduction of heat and electricity are promulgated as aspects of national energy use policy. In Western Europe district heating does compete with alternative sources, usually oil and electricity, but even here national policies encourage implementation of district heating if it is judged to be in the national and local interest.

The greatest rates of growth and most ambitious expansion
plans are found in Sweden, Denmark, Finland, West Germany,
Iceland, and the USSR. The USSR is the world leader in degree of penetration of district heating and extent of application of heat-electric coproduction. Eighty percent of the
urban heat demand is met by district heating and sixty percent of the heat is produced in conjunction with electricity.
Since 1955 the doubling period for heat sales from combined
power plants has been about six years. Expected developments
in this technology include use of nuclear plant reject heat,
operation of 1000 MWe combined heat-power plants, very long
distance one way transmission at temperatures up to 190°C [7].
Iceland has had geothermal district heating for some time.
Present service is to fifty percent of the population, and
work is underway to increase this to sixty-five percent. A
connection fee is levied and heat price is adjusted according
to gross sales. Total cost to the consumer is about twenty-
five percent less than imported oil [8]. Denmark has service
to one-third of all residents, and one-third of the heat is
produced in conjunction with electricity. Penetration to
forty percent or more by the early 1980's is expected, and
nuclear powered coproduction may eventually be used. It is
estimated that sixty-five percent of households along existing networks are connected [9]. This curious situation has
occurred because connection incentives were not given. This
may reflect historical conditions or it may reflect the
pricing structure in Denmark which favors oil in the very
short term, and district heating beyond a period of about
half a decade. Finland currently has extensive district
heating systems, reaching a total of 14 percent of all homes
in the country. The government plans additional connections
for a half-million people in Greater Helsinki alone within
the next 10 years, using waste heat from nuclear electric
plants. Sweden is currently meeting the heating needs of
about 20 percent of its population through district heating
systems. This includes several large cities such as
Vasteras and Malmo in which district heating supplies almost
the entire population. Based on growth of its nuclear power
program the Swedes project further growth of district heating
to an additional 1 million people, including the residents
of Stockholm, within the next 10 years. Ambitious plans include extension of service to low density areas and transmission of heat from a nuclear site at Forsmark to Stockholm--
a distance of 160 KM--predicated upon the development of
cheap, noncorrosive pipe materials. Overall goals call for
district heat service to all cities, the reject heat sources
being electric power plants and refuse incinerators. In West
Germany, between 7 and 8 percent of the population is now

served by district heating and studies of the feasibility of
service to the remaining markets are underway.

In all of these countries district heat has been cost
competitive with other heating methods. The conservation
potential and environmental improvement resulting from use of
district heating are so highly regarded as national priorities
that policies have been established which are favorable
toward planning and assessing or funding of district heating
systems. In some cases incentives are also supplied to induce consumers to connect to the system. In all cases district heating is strongly competitive with alternate energy
forms on a long-term basis without overt subsidization of
district heating authorities. Continued growth of the district heating industry bespeaks the benefits derived from it,
especially when perceived on a national level. Despite the
apparent decline of the U.S. industry, it is of great importance to re-examine the technical and economic aspects of
district heating for the United States in the context of
successful European technology.

Cost and Implementation Projections

We have analyzed the technical and economic aspects of
district heating implementation on two levels. The first
consists of a detailed examination of the demand profile and
analyses of model service systems to project district heat
costs for Washington, DC, Baltimore, Boston, and Philadelphia
[10]. The second level of effort is aimed at projecting maximal economic levels of implementation of district heating
for the nation as a whole [2, 11]. At the outset we assume
that all consumers contiguous to a system would use the service.

Urban Studies

The urban studies are based on actual floor space distributions, by building type, across each city. The demand
profile is calculated using local climates, floor space
distributions, and unit heating demand coefficients developed
by A.D. Little, Inc. [12]. A profile for Washington, DC is
shown in Figure 1. Data is presented on a square kilometer
grid. These are aggregated somewhat to form sections of area
6 to 10 KM^2 such that the demand is fairly uniform throughout a section and all the parts of a section are contiguous.
A separate distribution system is designed for each section.
A subtransmission network, such as shown in Figure 2, is used
to connect local power plants to the distribution networks.
Pipe sizes are selected for a design flow rate to meet 60
percent of the peak demand with water at 95°C input and 45°C

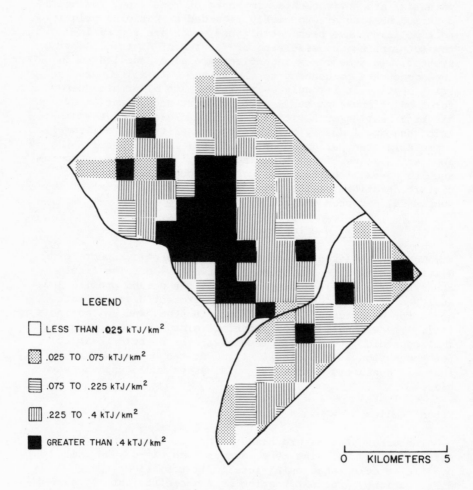

LEGEND

▢ LESS THAN .025 kTJ/km²

▦ .025 TO .075 kTJ/km²

▤ .075 TO .225 kTJ/km²

▥ .225 TO .4 kTJ/km²

■ GREATER THAN .4 kTJ/km²

0   KILOMETERS   5

Figure 1.   Heat Demand Profile, Washington, D.C.

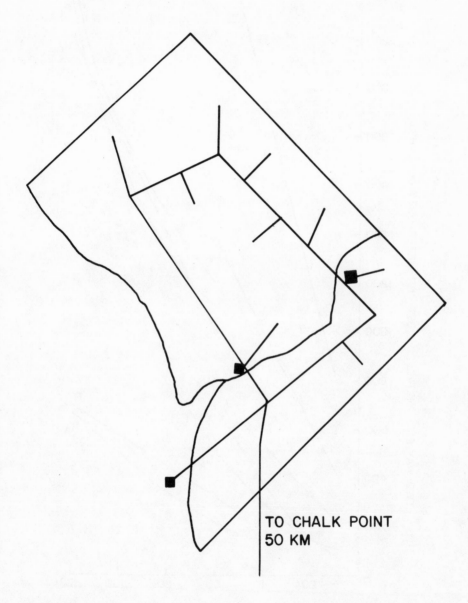

TO CHALK POINT 50 KM

Figure 2.   Model Sub-Transmission Network, Washington, D.C.

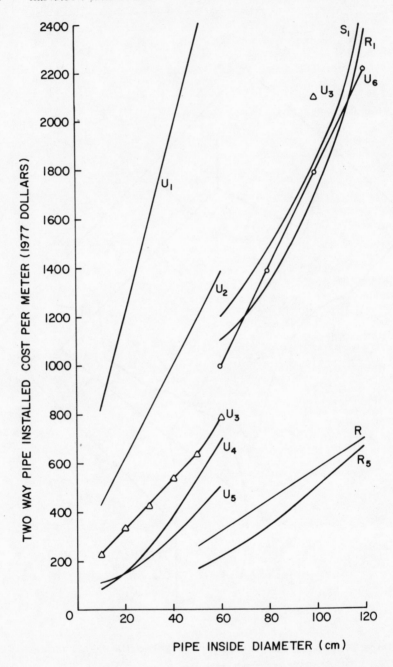

Figure 3.   Cost estimates for installation of district heating
            pipes. U, S and R apply to urban, suburban and
            rural installations, respectively.

return. Demand beyond this point could be met by raising the
water temperature or increasing the flow rate.

This approach toward small autonomous distribution subsystems is technically attractive for reliability and keeps
distribution pipe diameters small. In practice, small pipes
are much cheaper to install because of less material, easier
handling and less excavation requirement. This strategy is
also a realistic approach to load buildup of a district heating system because it allows operation of each section upon
completion.

The pipe system itself is carbon steel suitably insulated, encased in poured concrete and buried below the frost
line. Conservative pipe installation costs, shown as U_1, U_2
in Figure 3, are developed for each city on a per meter
basis in the spirit of small job construction, and include
current labor rates and detailed estimates of labor and
materials for <u>every</u> facet of the construction. U_1 reflects
construction under the most difficult conditions--crowded
intersections, for example--which are apt to be found only
within central business districts and comprise a few percent
of the construction on a city-wide basis. U_2 was used for
distribution and subtransmission and R_1 was used for rural
transmission lines from remotely sited plants.

Building retrofit and connection costs are estimated for
several building and original building equipment types in
detailed case studies. The potential heat recovery from
existing power plants made available through turbine modifications, and engineering assessment of the material, labor
and capital required to effect power plant retrofit are also
estimated in several detailed case studies. The estimated
loss in electric capacity on a retrofitted machine is
estimated to be about one-ninth of the useable heat gained
at $100^{\circ}C$.

Capital costs for distribution, connection and retrofit
are totaled for each section and the cost of heat at this
stage is computed on the basis of a fixed charge rate of 8
percent of capital and total section heat load. The lowest
density sections are bound to be marginal or uncompetitive
with conventional heat forms on their own merit. However,
on a system-wide basis each city is found to be quite competitive even when transmission, power plant retrofit, and
electric penalty are included. The principal results of this
part of the study are summarized in Table 1. A typical load
curve (Figure 4) shows that air-conditioning service could
greatly enhance the system load factor. We estimate the cost
of absorption AC equipment to be as great as the heating

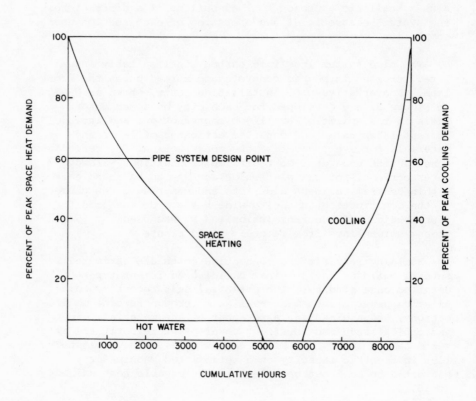

Figure 4.   Typical hourly load curve

Table 1. Summary of Urban Analyses

	PHILADELPHIA	BOSTON	BALTIMORE	WASHINGTON
-Estimated Population (1973-74, millions)	1.84	.62	.88	.73
-Heating Climate (C deg. days)	2834	3130	2586	2339
-Heating Demand				
Annual (10^{15}J)	79	32	41	30
Design Rate (GW)	5.9	2.3	3	2.2
-City Average Fuel Cost ($/GJ)	2.45	3.18	2.76	3.00
-Cost for Conventional Heat End Product ($/GJ)	5.08	5.87	4.98	5.08
-Estimate of Full Cost of District Heat City Average ($/GJ)	3.75	3.56	3.89	3.86
-Fossil Fuel Conservation (10^6 bbl oil equivalent, annual)	20	8.6	11	7.7
-Capital Cost Summary ($/kwt)				
Transmission	108	175	192	136
Distribution	200	185	208	191
Building Retrofit	120	102	124	128
Power Plant Retrofit	8-20	8-20	8-20	8-20
-Heat Available from Power Plants (GW)	6.5	4.2	6	6.8

connection and retrofit together. Higher inlet temperatures, which further reduce electric production efficiency, and greater flow rates are needed to meet air-conditioning demands with conventional absorption air-conditioning machines. In general we found this extra service to be of marginal economy.

In summary, these analyses reveal several interesting qualitative features. When regarded as a long-term investment, embodied in the low fixed charge rate, hot water district heat service in existing cities becomes an attractive option both for provision of long-term stable, comparatively low, heat costs and for affording reduction in heating fuel consumption by 75-80%. Also, contrary to widely held beliefs, a high load factor is not a requisite for the financial success of district heating. The lack of a high load factor, to be expected for many American cities,can be countered by a very high connection rate. Thus we conclude that rather than decry the lack of an ideal market, we should work to establish mechanisms for consumer acceptance of district heating.

National Assessment

The second stage of our study was concerned with projections of national levels of implementation of district heating within the economic limits commensurate with alternative energy sources, and the accompanying conservation of fossil fuels. For this task we used average values for degree-day data, and labor costs which were computed on the basis of a sample of all cities with a population of 100,000 or more [13]. This sample constituted a total population of 51 million (1970 Census). We used water main installation costs as reported by water authorities in several of the largest U.S. cities as a basis for estimating pipe installation costs [13]. There are two reasons why these are a good proxy. Water main installation is a well established procedure; and water authorities serve entire cities so that cost figures are typical. For piping material we selected polymer-lined polymer concrete insulated with rigid urethane. We used average labor rates and adjusted cold water main costs to accommodate two insulated pipes. Thus our approach was to construct typical installation costs on the basis of a well established industry and promising new materials. These estimates are shown in Figure 3 as U_5. U_1 and U_2 have already been discussed. U_3 are typical costs in Sweden for urban installation of steel pipes suspended within concrete conduit [5]. U_4 are estimates for steel pipe installation, with sandfill burial, in U.S. cities for conditions typical of garden apartment areas [14]. U_6 are average Scandinavian

costs for steel pipe suspended within concrete conduit [1].
S_1 and R_1 are suburban and rural estimates analogous to U_1.
R represents two and one-half times the typical rural costs
for installation of high strength carbon steel oil and gas
one way pipelines [15], and R_5 is the rural analog of U_5.
The credibility of U_5 is reinforced by estimates in [5] that
use of noncorroding materials could cut installation costs,
ie., U_3 or U_6 in half.

Average heat demand, ranging from 2 KW/person at highest
density to 2.6 KW/person at lowest density for residential
and commercial space and water heating, and subsystem capital
cost were than computed as a function of density of the
population served by a given system (Figure 5). Addition of
pumping energy costs and amortization of the capital costs
(at 10 percent per annum fixed charge) then yields the breakdown of heat cost as dependent on population density shown
in Figure 6.

The distribution subsystem includes all street mains and
city transmission lines, but no long distance transmission
lines from plants sited remotely from city limits. This
cost element was investigated and was found to depend very
strongly on the absolute demand, ie., population served by
the transmission line and climate, and also the length. For
typical distances of operating nuclear plants, 30 - 50 km,
cost contributions from $.25 to $1.00 per GJ were obtained.
The connection and retrofit subsystem includes service lines
from the street mains into each building, control valving and
heat exchanger equipment adapted to building size and form
of heat transport medium. Electrically heated buildings were
not considered, however. Retrofit in this case may significantly inflate the retrofit bill in some cities where electric heat holds a large share of the market, but on an overall system basis will have minor impact. The retrofit costs
were developed in case studies [10] of building and building
service types. Unit connection and retrofit costs, expressed
in dollars per square meter of building floor space, include
all materials and labor. These are $2.70 per m^2 for large
commercial and residential buildings, and $8 per m^2 for small
commercial and residential buildings. In our analysis we
alloted 35 m^2 and 10 m^2 of floor space per person for residential and commercial use, respectively. These are national
average figures. We classified low and high rise apartments
as large residential buildings and multi-family dwellings as
small residential buildings. Connection and retrofit cost
decreases with increasing population density (Figure 5 and 6)
because of the increase in proportion of apartment dwellers.

A surprising result is that modification of existing, or

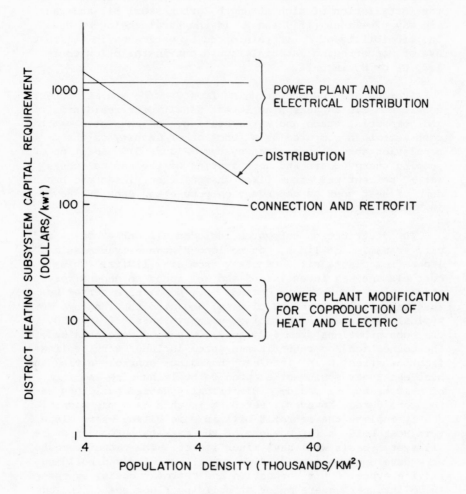

Figure 5.   Breakdown of capital costs for district heating
systems and comparison with electricity systems.
(The latter is expressed in dollars/KWe.)

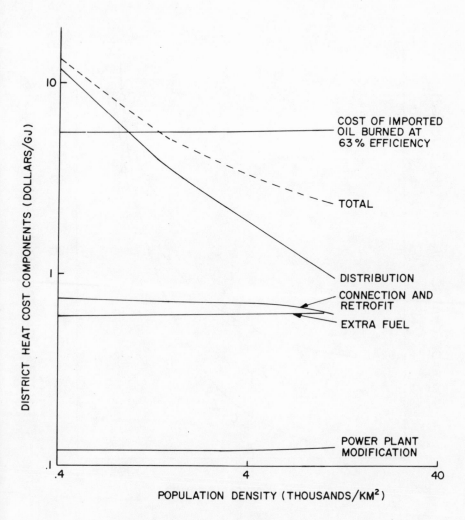

Figure 6.   Cost breakdown for district heat service and
comparison with oil heat

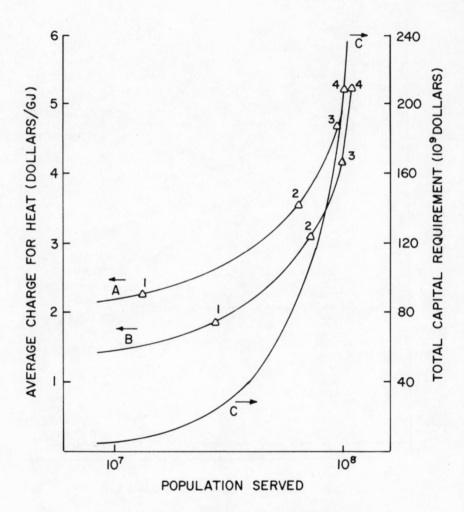

Figure 7.   Average cost of heat and capital expenditure
           projections for national implementation of district
           heating

new design, power plants, or industrial processes for that
matter, is a relatively very small cost element. A detailed
discussion of this factor and extra fuel is given in the
next section.

We projected levels of implementation by coupling this
profile to the population density studies of Haaland and
Heath [16]. Building retrofit may be included as part of the
district heating system and paid off in the same manner,
then curve A is obtained in Figure 7 which shows the weighted
average charge for heat costs less than or equal to the
effective energy costs of natural gas (1), imported oil (2),
and electricity (3), and total service level at an average
heat charge equal to the effective energy cost of imported
oil (4). Alternately, tax credits may be used to subsidize
any subsystems, in particular, to induce use of the service.
Thus, exclusion of connection and retrofit as a capital cost
raises the competitive serviceable levels as shown in curve
B, Figure 7. Use of such incentives is hardly capricious
since the long-term benefits to be derived from maximal
levels of implementation far exceed the initial investment.
Finally, curve C shows the estimated capital requirement at
service levels. Competitive implementation levels are
summarized in Table 2.

In summary, we estimate that between 50 and 55 percent
of the population (1970 Census) could be served by district
heating at average cost levels competitive with the current
energy price of imported oil. Total capital investment at

Table 2. Projections of National Service Levels
for District Heating

Service Levels (Millions)		
Full Amortization	With Connection Incentives	Competitive Cutoff
13.5	28	Natural Gas ($2.50/GJ)
65	72	Imported Oil ($5.25/GJ)
94	98	Electric ($10/GJ)
102	108	Imported Oil (Average)

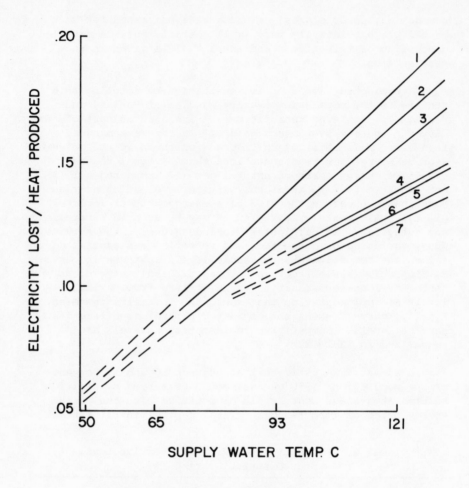

ONE STAGE EXTRACTION

1 - SMALL BACK PRESSURE TURBINE
2 - LARGE FOSSIL PLANT
3 - PWR

TWO STAGE

4 - 60 C RETURN
5 - 40 C RETURN

THREE STAGE

6 - 60 C RETURN
7 - 40 C RETURN

Figure 8.   Impact of low-grade heat production on power plant
electric production

these levels of service is estimated to be about $240 billion.
In other terms, this corresponds to $6,700 per household
(three persons average) connected, for a service that can
meet the full space and water heat demand of all residences
and commercial buildings. The initial cost estimate of $240
billion breaks down as follows: Three percent for power
plant modifications, twelve percent for service connection
and building retrofit, and 85 percent for the distribution
piping.

Power Plant Modification and Heat Availability

The amount of heat that can be extracted, as a function
of temperature, from plants designed for coproduction of
heat and electric has been well studied. Figure 8 [10]
shows the ratio of electric capacity lost to heat output for
various extraction strategies. To produce water at about
100°C under the best of conditions from a single turbine unit
would require a loss ratio of about 0.1. This reduces electrical generating efficiency from 33 percent down to 25 percent. Under this condition, more plants would be operated to
meet the same electric demand and would burn additional fuel
equal to about 40 percent of the heat supplied. This incremental cost ranges widely depending on the fuel type.
Nationally it amounts to about $.60 per GJ of heat generated.
The incremental costs incurred by modification of a plant in
the design stage fall into two categories: turbine modifications and balance of plant. The former may include
addition of several extraction ports and incorporation of high
backpressure operating features for optimal use of fuel. The
latter feature is readily available and is a very small cost
item, but machines with both features are not presently
available from American manufacturers. The balance of plant
incremental costs arise from additional piping, heat exchangers and controls, which in case studies [10] have been
estimated at about $14 per original design kW(e), installed.

Retrofit of existing plants is a more delicate problem
and has not been investigated generally. The strategy that
would or could be applied to any machine depends on many
factors. Thus, there exists a wide range of possibilities.
We have considered two extreme cases [10].

The first strategy is to extract steam at the crossover
point between intermediate (I-P) and low pressure (L-P) turbine units, and to operate the L-P at maximum recommended
condenser backpressure, typically 125 mm Hg (absolute).
Additional equipment amounts to some extra piping, a heat
exchanger and controls. The actual modifications could be

Table 3. Technical and Economic Assessment of Power Plant
Modification for Heat-Electric Coproduction

	New Design	First Retrofit Strategy	Second Retrofit Strategy
Turbine Modification $/KWt	3	1	9
Balance of Plant Modification $/KWt	8.7	7.5	8.3
Temperature of District Heating Water	100°C	57°C (1) 60°C (2) 63°C (3)	100°C
Heat Gained / Electric Lost	10	12 (1) 9 (2) 7.8 (3)	9
Electric Production Efficiency	25%	30% (1) 28.5% (2) 26.7% (3)	25%
Energy Conversion Efficiency*	85%	70% (1) 70% (2) 69% (3)	79%

*   Without use of stack gas heat (Nuclear Plant Yields 95%)
(1)   7% of throttle flow extracted at crossover
(2)   12% of throttle flow extracted at crossover
(3)   15% of throttle flow extracted at crossover

effected in a short time and would not shut down the plant
beyond a normal maintenance period.

At the other extreme is radical re-design of the L-P
unit. This strategy includes addition of two large capacity
extraction points and subsequent operation of the unit at
125 mm Hg (absolute). This strategy is more costly than the
first but yields hotter water (Table 3). In this case,
greater heat exchange area, more piping, and more complex

control systems are required. The LP case and rotor must be removed for extensive modification estimated to be almost as costly as a new unit. In addition, a plant must be shut down for an interval in excess of the typical overhaul period. The loss in productivity is a relatively small, one time only, cost item.

Both strategies allow operability of the turbine at full electric capacity. The former is best applied to small, older units which generally are found several to a plant. In this case, one unit could furnish a base load supply as indicated in Table 3, while the others could supply cross-over steam in series to drive the temperature up as needed. The latter strategy is best applied to large, newer, single unit plants. Adjustment of district heat flow rates can be had by varying the relative amounts of extracted steam. Utilization of stack gas heat would increase the heat rate and temperature and reduce the amount of extra fuel burned without affecting the generating efficiency. A summary of power plant analyses appears as Table 3. These analyses suggest that power plant modification would cost between $8 and $20 per kw(t), and in consideration of the yearly temporal variation in district heat flow rate and temperature sent out (see Figures 4 and 8), an average 10 units of heat per unit of electric lost represents a conservative basis by which to estimate extra fuel consumption.

Conservation Potential

The potential oil and gas conservation, expressed in terms of oil equivalent and shown in Figure 9, reflects that fuel which would have to be burned employing conventional technology to furnish space and water heat less the additional fuel required by power plants. The gas that is liberated from these applications can be re-directed into other areas where oil is presently used to essentially eliminate the use of oil for space and water heating. District heat service to half the population could reduce imports by about 40 percent. The payback period shown in Figure 9 is the interval required to recover the capital investment based on the face value of reduced imports, thus reflects the direct cost of imported oil. At service to half the population this would occur after 18 full years of operation, at present import prices.

Other Benefits

The key variables in our analysis are fixed charge rate and availability of cheap low-grade heat. Ten percent per

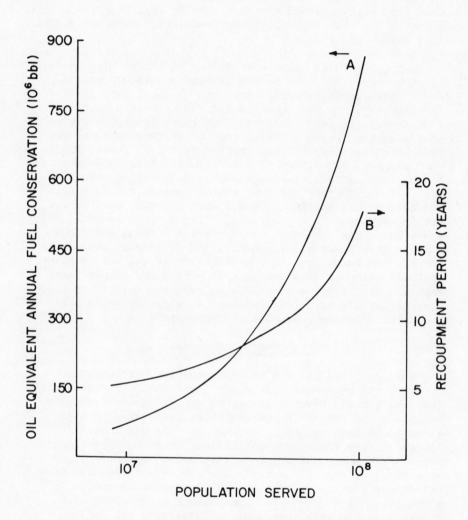

Figure 9.   Conservation and payback projections for national
implementation of district heating

annum, not uncommon for municipal systems in Europe, is probably too low to attract private capital investment. More complex funding schemes may be used in Europe but to the same end result. Clearly as the fixed charge rate increases, maximal service levels decrease. If primary energy is used to furnish district heat, this cost element could easily increase fivefold, thereby considerably reducing maximal service levels. It is also obvious that significant conservation can only be realized if secondary heat sources are used. These are in ample supply.

Granted our assumptions, many other benefits become apparent. Of greatest importance to consumers, and a useful lever in encouraging connection, is that long-term heat costs are fairly well determined at the outset because district heating is capital intensive. Secondary heat cost is quite insensitive to primary fuel price [2] and district heating system operation and maintenance costs can be very low. The availability of secondary heat should be more certain than that of conventional fuels now used for heating since the primary processes which produce the secondary heat are more adaptable to fuel changes. Thus district heat supply should be more reliable than conventional service in the long term. The impact of benefits perceived at a regional and national level is more difficult to quantify. Improvements in air and water quality are aesthetically appealing, but the health benefits, recognized for some time [4], could be far reaching. The potential for employment during construction of district heating systems is very great, not only for the construction and building trade industries but also for designers, planners, managers, and operating and maintenance personnel. The infusion of large amounts of capital into hard pressed cities to produce this useful technology is a constructive means of aiding cities. Disuse of oil and gas by the heating sector increases the availability of these as industrial feedstocks and smooths out seasonal variations in energy flows, thus enabling reductions in storage capacity and peak demand capability of oil and gas suppliers. Our estimate of the maximal annual conservation potential is equivalent in dollar terms, at present world oil prices, to half the 1977 U.S. trade deficit of $27 billion. Reduction of consumption by this amount would constitute a significant step toward energy efficiency, and improve our nation's image abroad. An additional benefit is that by supplying district heat from coal and nuclear power plants these fuels can directly substitute in the space and water heating market without recourse to synfluid or additional electricity production.

Institutional Issues

Recent studies [10, 14, 17, 18] conclude that district
heating would be technically feasible and economically competitive in many cities. We suggest here that developments
in pipe technology, which reduce installation cost and increase longevity of pipe systems, could increase the
feasibility of district heat service to half the population.
However, any enterprise for reducing the amount of fossil
fuel used for space and water heating will encounter difficulties of a non-technical nature. District heating en-
cumbers a great number in this country just because of the
existing institutional structures.

The degree of implementation of district heating in
Europe appears to be in direct proportion to the degree of
involvement of the government in providing policies regarding
energy use. The same rules apply in the U.S. It will have
to establish policies that will overcome obstacles in a
manner acceptable to private and public interests while
implementing desirable new technologies. It will be necessary for the government to support assessment and planning,
to introduce mechanisms for long term low cost funding and
perhaps institute regional authorities to manage systems
which may cut across political boundaries. Furthermore,
incentives may have to be provided to consumers to encourage
connection and also to industry to encourage development of
secondary sources of heat. Long-term government loans could
certainly be provided on the basis of it being more beneficial to invest the money internally.

District heating would surely compete with conventional
service industries, and eradicate many heating oil dis-
tributorships and considerably reduce natural gas sales in
urban areas. District heating may become a new utility,
separate from those now in existence, especially if developed
municipally, or it may represent a growth area for existing
utilities. In any case ownership, licensing and regulation
must be clarified. This includes definition of system
reliability criteria and the liability of industries to
supply heat upon demand, and establishment of criteria for
cost allocation particularly between electric rates and heat
rates. This may be further complicated by the growing
practice of power pooling and economic dispatch of power,
since heat demand may necessitate operation of an inefficient
plant.

To effect satisfactory economics from district heating,
a very high percentage of potential customers in each market
must use the service. Promises of reliability and long-term

cost advantages may not be sufficient to ensure market penetration. Suitable tax policies rewarding the use of district heat service and discouraging the use of conventional sources may encourage connection. A tax structure that encouraged full use of the energy content of high-grade fuels could further encourage industries to become secondary heat sources. Confidence in the service may be developed in part by actual demonstration projects in several cities.

The feasibility of placing new service lines in city streets can only be determined by close scrutiny on a local level. It is acknowledged that in some cases this can only be done with great difficulty, at great expense. However, there is no reason why pipes have to be installed in streets. It is possible that the space under sidewalks is less cluttered than the adjoining street and construction in this way would interfere less with traffic and be more amenable to maintenance service. There is also ample precedent for passing service pipes through building foundations and by-passing excavation altogether. Construction of this type would be considerably cheaper, but require access rights for repairs and maintenance. Away from urban cores it may be possible to install service lines through customers' backyards and reduce costs to such a low level that even low density suburban service becomes competitive.

Summary

We conclude that district energy systems are a very attractive option for meeting future U.S. energy needs. The technology is well established. The cost/benefit yield is favorable, and the conservation potential is significant. Application of district energy should occur in urban and densely populated suburban areas. The remaining portion of the space conditioning and domestic water heat demand located in rural and low population density communities appear to be better suited to other forms of system substitution.

References

1. Charles F. Meyer and Walter Hausz, "Role of the Heat Storage Well in Future U.S. Energy Systems," General Electric Company - Tempo Center for Advanced Studies, Santa Barbara, CA, Ge76TMP-27.

2. John Karkheck, James Powell, and Edward Beardsworth, "Prospects for District Heating in the United States," SCIENCE, March 11, 1977, Vol. 195, pp. 948-955.

3. Patricia A. Kelsey, "European, Asian Countries Describe
    District Heating-Cooling Systems, Hardware, Metering,"
    Air Conditioning, Heating, and Refrigeration News,
    July 14, 1975, p. 1, 16-19.

4. A. E. Haseler, "District Heating in New Cities," Journal
    of the Institute of Heating and Ventilating Engineers,
    September 1965, pp. 180-196.

5. Kjell Larson, "District Heating: Swedish Experience of
    an Energy Efficient Concept," from the Swedish Approach
    to Current and Future Energy Issues, 1977.

6. "District Heating in the Union of Soviet Socialist
    Republics", International District Heating Association,
    Pittsburgh, 1967.

7. Y. Y. Sokolov, "Current State of the Introduction of
    District Heating Systems in the USSR and the Prospects
    for their Development During 1971-1975," Heat Transfer
    Soviet Research V6 No. 2, 1974, pp. 1-9.

8. Paul Lienau, "Reykjavik District Heating," Geo Heat
    Utilization Center Quarterly Bulletin, OIT, Klamath
    Falls, Oregon, August, 1977.

9. W. Mikkelsen, "Development of District Heating Systems
    in Denmark," Bruun and Sorensen A/S, Arhus, Denmark.

10. J. Karkheck, Ed., "Technical and Economic Aspects of the
    Implementation of District Heating in Four Northeast
    Cities," Brookhaven National Laboratory Report - in
    preparation.

11. J. Karkheck and J. Powell, "Waste Heat as an Alternative
    Energy Source", Brookhaven National Laboratory Report
    BNL-23751, December 1977.

12. A. D. Little, Inc., MUPP Study.

13. J. Karkheck, E. Beardsworth, J. R. Powell, "The Technical
    and Economic Feasibility of U.S. District Heating
    Systems Using Waste Heat from Fusion Reactors,"
    Brookhaven National Laboratory Report BNL 50516, February
    1976.

14. M. Olszewski, "Preliminary Investigation of the Thermal
    Energy Grid Concept," ORNL/TM-5786, Oak Ridge National
    Laboratory, Oak Ridge, TN October 1977.

15. Pipeline Economics, The Oil and Gas Journal, August 22, 1977.

16. C. M. Haaland and M. T. Heath, Demography 11, No. 2, 321, May 1974.

17. Potential for Scarce Fuel Savings in the Residential/ Commercial Sector through the Application of District Heating Schemes, (working paper), Argonne National Laboratory, Argonne, IL, 1977.

18. C. L. McDonald, "An Evaluation of the Potential for District Heating in the United States," Batelle Pacific Northwest Laboratories, BNWL-SA-6259, November 1977.

Acknowledgment

The work about which this chapter was written was performed under the auspices of the Department of Energy.

COMMENTS

Gary Kah (Donovan, Hamester & Rathen, Inc.): Although it is a convenient measure, it is important to realize that the heating-degree day is technology-dependent and is an outmoded measure in the face of high energy prices. No extra heat energy (beyond internal dissipation) is required down to $65^{\circ}F$ to maintain a 70°F internal temperature. But a well-designed passive-solar, fully insulated and weather-stripped home may not require any external thermal energy until the outside temperature drops to the 40's and 50's Fahrenheit on the average. The author has therefore overstated the possible benefits of district heating by as much as 40%, if such design features were implemented, as many may be.

Author's Reply: Implementation of passive solar features in existing homes would provide substantial energy savings, but probably not practical in the majority of cases. Improvements in insulation and weather stripping are more feasible for existing homes, but will not substantially alter either the heating requirements or our conclusions regarding the benefits. The existing housing stock will be in place for many decades, and must be serviced by either continuing to supply even more scarce fossil fuels, by solar collectors or by district heat. We feel the latter to be the best solution for areas with relatively high population densities.

Robert Bingham (Philadelphia, Pa): The authors' system service territories (based on population densities) defined the black ghetto of Philadelphia and the District of Columbia. This should convince us very quickly that there are immense social problems in connection with such large-scale projects and suggests that more serious considerations be given to them before going further.

Author's Reply: It is true that district heat is more viable in densely populated areas, and that, in general, such areas tend to be connected with low income groups and/or ghettos. This should be regarded as an opportunity rather than a problem, however. The construction of district heat systems in such areas will provide many local construction jobs, as well as reduce the amount of money that low income groups must presently pay for heat.

David K. McGuire (Upsala College): Could the authors give more details about long distance transmission of hot water (for example, 100 kilometer transmission of nuclear waste heat at 90°C)? What problems might arise from a sudden shut-down of the nuclear plant?

Author's Reply: Long distance transmission is technically feasible and is practical now. The economic practicality of

long distance transmission primarily depends on the transmitted heat demand. In general, long distance (~100 km) transmission requires heat demands on the order of 1000 MW (th);
lower heat demands are only economical at shorter distances.
The question of power plant shutdown is not unique to nuclear
plants - any heat source will require backup by reserve heat
sources and transmission lines. Nuclear plant reliability
is, in general, as good as that for fossil power plants or
industrial waste heat sources.

4

A Review of the
HUD Total Energy Experience
and the MIUS Program

Jerome H. Rothenberg

Abstract

In the early 1970's, the Department of Housing and
Urban Development (HUD) undertook a program to develop
utility delivery mechanisims for housing. One specific
thrust of this program is the integration of power generation
and environmental conditioning - Total Energy - in the
multifamily sector. The National Bureau of Standards
performed a feasibility survey of HUD Operation Breakthrough
sites to determine which sites would be suitable for
implementation of a highly instrumented total energy system.
The Summit Plaza Development at Jersey City, New Jersey, was
selected. The site consists of 486 residential units, plus
other space. The plant provides electricity, hot water for
space heating and domestic hot water, and cold water for air
conditioning. The site occupancy started in 1974, and the
total energy plant was operating then. In addition, the
site includes a Pneumatic Trash Collection (PTC) System
which collects all solid waste from the site. The site was
heavily instrumented and data collected and reported for a
one-year period from November 1975 to October 1976. Both
technical and economic data are collected. This data and
the plant analysis will be reported.

Background

HUD's integrated utility activities can be broadly
divided into two programatic areas; the Total Energy Plant
at Jersey City, New Jersey, and the MIUS Program. While
different in many ways, these program activities draw on a
common conceptual base, and together form the basis of
upcoming HUD activities. The chronology of each program
area is reviewed below as a prologue to the next phase of
this research activity.

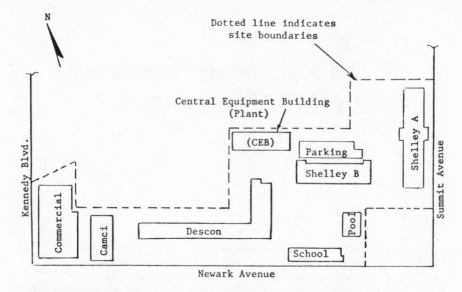

Figure 1.   Relative location of individual buildings at the
            Jersey City Total Energy Site

Figure 2.   Aerial view of the Jersey City Total Energy Site.
            See above to identify the buildings at the Total
            Energy Site

Figure 3.   Central Equipment Building which houses the Total
Energy Plant. The five engine-generators are located just inside the six sets of doors. The central tower houses the cooling towers. Ventilation
air for the plant is supplied through the central
grill.

HUD Total Energy Research

The conventional generation of electricity in larger central power plants usually wastes approximately two-thirds of the energy available in the fuel input to the plant. The largest fraction of this wasted energy takes the form of heat released to either the air or water. In addition, there are losses due to the transmission of electric power from the central plant to the local user. In the past, these wastes have been tolerated because of the apparent higher reliability of such interconnected central systems, lower capital and operation costs of larger central systems, and the higher "quality" of the electrical product.

With the increased recognition that the United States cannot continue to waste energy by dissipating it to the environment, increased research has been directed toward the capture and useful application of this waste heat or co-generation. There are two basic approaches to this problem. First is the direct use of central plant waste heat to useful purposes through district heating and cooling systems. Second is the utilization of smaller generating plants, located at the point of electrical use, wherein the waste heat of generation is used to satisfy local thermal needs. This latter approach has been successfully applied by industrial cogenerators[1], in captive markets[2], and in selected residential and commercial applications. On-site power generation has been the focus of the HUD research at Jersey City, New Jersey.

One of the goals of the Department of Housing and Urban Development is to assist sound development of the Nation's communities. HUD's goals include "decent housing and a suitable living environment for every American family" at a cost they can afford. Space conditioning, power and domestic hot water are elements of a decent home, and the

---

1   Cogeneration as used here is the simultaneous production of electricity and termal energy (steam or hot water).

2   Captive markets refer to those applications such as hospitals, TV stations, and others where there is a requirement for on-site generation as a backup to the utility grid. In such applications, TE is economically viable based on the smaller incremental cost of expanding the building heating and cooling systems and the required backup generation capacity.

requirement for these services at reasonable costs demands
that consideration be given to application of efficient
techniques for the provision of these services. This
purpose was in mind when HUD began to examine the Total
Energy concept and its application to residential housing
in 1970 when the Operation Breakthrough sites were in the
planning process.

One of HUD's initial conclusions was that there was
insufficient operational information available to determine
how effectively the TE concept could be applied in the
residential sector. Therefore, the Office of Policy Development and Research (PD&R) of HUD, funded the construction
and instrumentation of a Total Energy Plant at an Operation
Breakthrough (OB) site in Jersey City, New Jersey.

Site selection was based on the conclusions of a
National Bureau of Standards (NBS) study of the feasibility
of the application of TE at OB sites. The plant was
designed by Gamze, Korobkin and Caloger (GKC), Inc. of
Chicago, and constructed with HUD funds. The site is
presently owned by the Starrett Housing Corporation, New
York City, and under a six year agreement between HUD and
Starrett, HUD continues to operate the plant; through a
contract with GKC. NBS, with HUD funding, designed and
installed the entire data acquisition and processing
system which it continues to operate. The data obtained
from this system is the basis of a plant performance report
to be released in 1978. An Interim Performance Analysis
was published in 1977.

The site itself consists of 486 dwelling units in six
mid and high rise buildings, 50,000 square feet of commercial space, a school and a swimming pool on a 6½ acre site.
Electric and thermal power is provided from the TE plant
which includes five 600 KW diesel engine-generators, two
4.0 MW fire tube boilers, and two 546 ton absorption
chillers.

The plant was put into operation in 1974, and an
analysis of one year of operation by NBS determined that a
conventional central heating and cooling plant, purchasing
power from the local electric utility would have consumed
17.3% more resource energy. This performance was achieved
at costs approximately equal to that of "conventional"
systems and with a reliability of 99.8%. A review of the
first year of acceptable data indicated the need for some
modifications to the plant, which, where feasible, have
been or are being implemented.

Table 1.

HUD UTILITIES DEMONSTRATION SERIES

Vol. No.	Title	Organization	Availability	NTIS No.
One	Survey of User Acceptance of the Solid Waste Removal Systems at Operation Breakthrough Sites	Hittman	-	PB 257-474
Two	All Sites (Evaluation of Refuse Management Systems of the Operation Breakthrough Sites)	Hittman	HUD	PB 260-495
Three	Evaluation of the Refuse Management Systems at the Jersey City Operation Breakthrough Site	Hittman	-	PB 280-551
Four	Executive Summary	Hittman	-	PB 280-143
Seven	Performance Analysis of Jersey City -- Interim Report	NBS	NBSIR 77-1243	PB 269-517
Eleven	Engine Performance Report	NBS	NBSIR 77-1207	PB 264-427
Twelve	Jersey City Design Report	GKC	-	-
Fourteen	Engine Generator Emissions	NBS/YORK	NBSGCR 77-104	PB 276-102
Sixteen	J/C Incinerator Feasibility Study	American Hydrotherm	-	-

The entire HUD utility experience is being documented in a special series of reports entitled "Utilities Demonstration Series," of which several documents are presently available (Table 1).

MIUS Concept

The MIUS (Modular Integrated Utility Systems) concept is a step beyond Total Energy. In addition to providing electricity and space conditioning, the MIUS provides solid waste processing, liquid waste processing, and/or potable water supply. In the MIUS concept, these utility services are provided on site in an integrated system. For example, heat from the incinerator can replace the use of a boiler, or wastewater treatment subsystem effluent can be utilized for cooling tower make up water for the power generated subsystem.

Integration also extends into the operation of the MIUS wherein all of the utility systems can be controlled from a single location, as compared to the individual operation of conventional non-integrated utility subsystems. Also, the system performance can be improved through integration, with each subsystem probably operating at a more efficient energy level.

Finally, the MIUS concept includes modularity, a call for the design and construction of a utility system on-site in steps or modules consistent with the growth of the community it serves.

To summarize, the MIUS concept is the integration of on-site power generation and space conditioning with potable water, solid waste management, or wastewater management subsystems, in a manner that reduces the total cost of utility service and ameliorates the adverse environmental impact providing these services.

The MIUS Program

The MIUS Program consists of three phases; a technology investigation and assessment phase, a demonstration phase, and a dissemination phase. Phase I called for investigations and assessments of each of the technological components of MIUS. Working with HUD, NASA, ORNL and NBS have prepared a large number of reports analyzing the performance and characteristics of the MIUS concept, the various subsystems, and the MIUS components. In addition, various analytic tools such as the ESOP computer program were developed. A test facility[1] was developed and operated by NASA. This phase of the program is documented in more than thirty publications including a Generic MIUS Environmental Assessment and a MIUS Technology Assessment in two volumes.

Phase II of the MIUS Program called for the demonstration of a MIUS. The intent of the demonstration was to have a private sector entity design the MIUS and address institutional barriers as they were encountered. HUD's objective was to determine the validity of concerns about institutional barriers, and to document the solution to such problems thus providing a roadmap for future MIUS owners.

As a result of a competition, an award was made to the Interstate Land Development (ILD) Corporation for the design of a MIUS at St. Charles, Maryland. HUD allowed the developer a free hand in the MIUS design, only requiring that the developer be cost effective in satisfying a site specific performance specification. The design process continued through September 1977, and the design and institutional reports are being developed by ILD for release in 1978.

The demonstration satisfied several of the institutional goals of HUD. ILD, with the local utility, arrived at an arrangement by which both organizations could profit from a MIUS interconnected to the grid. In addition, the Maryland Public Service Commission ruled this MIUS to be exempt from power plant siting laws and state legislation.

The third phase of the MIUS Program is directed at dissemination of the MIUS information developed in Phase I and Phase II. Specifically, the activity planned will use workshops and seminars, with design guides all intended to provide MIUS information to key decision makers, including design evaluation techniques; as well as site specific concept analyses and studies.

Phase III has taken shape over the past few months. We expect to have contractors at work soon and with luck to initiate workshops by Spring.

---

1   The MIST test facility continues to function at the LBJ Space Center under a HUD and DOE cooperative arrangement.

A key component of Phase III is the existence of a real life test at Jersey City. As pointed out above, the addition of the incinerator with heat recovery to the TE plant provides the first MIUS. It will provide technical and economic data on the concept and be a basis of ongoing research.

5

Economic Aspects of Choosing Comfort-Conditioning Modes

Robert Thomas Crow

Comfort conditioning for residential and commercial buildings accounts for approximately 15% of total U.S. end-use consumption. Space heating, in particular, has become a technological olympiad in which the competition is fierce among fossil-fuel fired forced-air furnaces, electric furnaces, and radiant electric heating. New entrants to the competition include heat pumps and, most recently, solar heating units.

It is impossible to make meaningful <u>general</u> assertions about the relative superiority of one type of system or another because the appropriate choices depend upon such factors as climate, type of structure, type of housing installation (new or retrofit), relative and absolute prices of different forms of energy, and consumer tastes. The objective of this paper is to discuss these factors and examine how public and private choices relate to comfort conditioning from the perspective of economic analysis.

1. Comfort-Conditioning Efficiency

"Efficiency" is perhaps the most misused and over used term -- other than "crisis" -- in the entire field of energy analysis. In discussing efficiency, one must first distinguish between thermodynamic efficiency and economic efficiency -- two concepts that often have entirely different private and social implications when applied to energy problems. Further, even if it is clearly understood that thermodynamic efficiency is the relevant frame of reference, one must distinguish between "first-law" efficiency and "second-law" efficiency. Considering the distinctions I wish to make,

---

*The views expressed in this paper are the sole responsibility of the author. No endorsement of these views by the Electric Power Research Institute has been either expressed or implied.

however, the choice of thermodynamic measure is immaterial.
Either one, if pursued with single-minded dedication, will
lead to solutions that are privately and socially absurd.
The reason for this is that these physical measures implicitly treat energy as the sole scarce resource whose use is to
be optimized.

Inherent in a single-minded focus on thermodynamics are
two critical errors. One is the emphasis on energy to the
exclusion of all other productive factors. Energy is not the
only resource that is scarce. Other resources that must be
considered in the optimal use of an energy system are the
capital, labor, materials and energy used in manufacturing
the end-use equipment (in our instance, comfort-conditioning
equipment) and the capital, labor, materials and energy that
are used in making the energy of a particular fuel resource
available. An example of the error of assuming that energy
is the only scarce resource would be the diversion of highly
skilled labor, specialized equipment,and expensive non energy
materials to the task of saving an amount of energy so small
that the monetary value of the energy savings would not offset the cost of the other productive factors. The second
major error in treating energy as a uniquely scarce resource
is that it is not scarce per se. What is scarce is fuel that
can be converted to useful heat or work without incurring too
high a penalty in the use of other scarce resources. It is a
mistake to consider "energy" as a homogeneous entity. The
focus should be shifted from energy in general to those specific materials that are readily transformable to energy.
This leads to the suggestion that some reasonably comprehensive common measure of value be adopted for all scarce factors of production so that informed trade-offs can be made.
Money has served this purpose in many contexts for several
thousand years.

Considering energy essentially as a single, undiffer-
entiated resource ignores the fact that there are many potential sources of energy that current engineering could transform into desired forms, given the appropriate amount of
energy and nonenergy inputs. Relatively few of these sources
are used extensively, however, because only a few (petroleum,
natural gas, coal, nuclear fuels, and hydropower) require
relatively little (compared to other possible sources) in the
way of energy and nonenergy inputs (as reflected in monetary
costs) to transform them to desirable forms of energy.
Furthermore, even among this relatively small group of readily transformable sources, there is wide diversity in the
resources needed to make them available and in the ease of
application at the point of use. From this perspective, if
one has abundant coal that is easily available and small

reserves of natural gas, it is not necessarily folly to use coal whenever possible, even if the thermodynamic efficiency of natural gas is higher in a given application. Certainly, if solar radiation could be converted to usable energy without a massive infusion of capital, labor, and other materials, one would use it regardless of the efficiency of the conversion process because it is totally renewable. Thus, economic efficiency should be considered with, or in place of, thermodynamic efficiency in evaluating the efficacy of one form of energy or another in a given application.

Economic efficiency with respect to energy may be defined as the sum of the private and social values of a given service provided by energy relative to the sum of the private and social costs of making that energy available. Note that economic efficiency allows for "externalites", such as the social costs due to air pollution in the combustion of fossil fuels. It also includes room for qualitative differentiation in the performance of essentially the same service. For example, the most important aspect of an automobile is that it is a passenger-operated vehicle that will allow one to go from point A to point B. Even though all automobiles perform this service, there is little question that large, luxury automobiles offer a qualitatively different performance of this service than subcompacts. Finally, the definition of economic efficiency allows for the possibility that energy resources are not priced at their true values.

In addition to the distinction between thermodynamic and economic efficiency, a second important distinction is the difference between the efficiency of components and the efficiency of a total system. For example, the efficiency with which a furnace produces heat from energy input is only part of the story. Also to be considered in the system's overall performance are the energy requirements of the blower fan, duct losses in unconditioned spaces, heat losses due to infiltration through exhaust stacks (where relevant), and losses through continuously burning pilot lights (where relevant). In addition to the production and distribution elements of a particular mode of heating, the total system also includes heat gains from other energy-using equipment and the construction characteristics of the building being conditioned.

2. Consumer Choice Aspects of Comfort Conditioning

In the end, aside from legislation to deliberately restrict choice, the comfort-conditioning systems that will be relevant are the ones that are chosen directly or indirectly by those who will use the systems. Thermodynamic efficiency

thus becomes important only to the extent that it is relevant
to the production of desired services or to the dictates of
government. I submit that thermodynamic efficiency considerations do not typically play a major role in the decisions
of households and firms. Economic efficiency would seem to
be much more relevant to the choices of consumers, since it
balances the values of <u>all</u> scarce factors. In a market without government intervention, choice will be based simply upon
an optimization of private gains and private sacrifices.
Government intervenes in market choices largely because the
private interests are not congruent with those of society as
a whole. Thus, social costs and gains also become relevant.

Social aspects consist essentially of two types of
actions: direct actions to restrict choice, such as barring
installation of a particular mode of comfort conditioning;
and the manipulation of prices through taxes or subsidies, in
order to ensure that energy and other resources will be
bought and sold at prices that closely represent their social
values. The focus of this paper is directed at the choice
aspects of comfort conditioning, so it will include taxes or
subsidies as an element of price and exclude restrictions on
choice.

The most readily identifiable aspect of consumer choice
is the interplay between consumption, ability to pay (be it
through income or wealth), and prices. However, in dealing
with consumption decisions among closely competing forms of
goods and services (such as different modes of space heating)
consumer choice analysis has also been concerned with non-
price characteristics. However, even though we know that the
qualitative choice aspects of comfort conditioning may be
of critical importance in determining the success of new and
existing technologies, we now know very little about how
people react to them. Even the traditional aspects of economic analysis -- considerations of price and ability to
pay -- are complicated in considering comfort-conditioning
systems. As in the case of an automobile and its services,
or other types of capital equipment and related services, we
may distinguish between investment costs, costs of installation, and costs of operation and maintenance. If we were to
conduct our investigation in a context in which pricing,
financing, and consumer choice decisions were made in a
completely rational and efficient context, all of these different costs could be rolled into a neat bundle, such as life
cycle costs. However, it is not at all clear that things are
that simple.

Consider a residential comfort-conditioning choice. In
the vast majority of cases, the selection of a comfort-

conditioning system is a <u>fait accompli</u>. It simply comes with
the house, except in those relatively rare cases in which a
house is custom-built. Thus, the choice aspect is limited to
whether to keep a comfort-conditioning system or replace it.
An important factor is the builder, whose motive is to build
a house that will yield the greatest profit. It is only by
happenstance that heating and cooling systems selected under
this motive will be the most efficient from the point of view
of thermodynamic efficiency or economic efficiency. (A nota-
ble exception is the case in which the efficiency of an
energy system becomes a major marketing aspect of selling the
house.) Thus, the investment and installation costs of a
particular space-conditioning system are hidden in the cost
of construction and are only one aspect of the total bundle
of characteristics that the house provides. This means that
comfort-conditioning life-cycle costs, at least from the
occupant's point of view, are extremely difficult to calculate and are only of marginal relevance.

What will be important, of course, will be the percep-
tion of operating and maintenance costs for the period in
which the occupant resides in the building. Occupancy again
raises a difficult issue in a society in which places of
residence and places of business change fairly frequently.
It may be difficult for occupants to capitalize on an investment in an energy-efficient system. If energy-saving investments cannot be capitalized when the building is sold, the
energy savings realized during the terms of occupancy may not
be worth incurring the required additional investment and
installation costs.

Thus, even traditional aspects of economic analysis are
rather difficult to pin down. A complete model of the choice
of comfort-conditioning modes would appear to rely heavily
upon a characterization of the builder's choice as well as
the occupant's choice. At the very least, since it is not at
all clear that individual consumers maintain private discount
rates that match the market interest rate, a prudent approach
to understanding heating-system choice would attempt to
separate investment and installation costs from operating and
maintenance costs, thus avoiding the seductive simplicity of
using total life-cycle costs as a single criterion.

Turning now to the conceptionally more difficult question of service quality, the qualitative component does not
appear to be as important as it is in may other kinds of
energy-using goods, such as automobiles. On the other hand,
even though we believe qualitative aspects to be of less
importance for comfort-conditioning systems than for many
other energy-using systems, they are probably not negligible,

and we know very little about quality variation between comfort-conditioning modes and how consumers react to the
variation.

A clue to what characteristics might prove to be most
important is afforded by a recent study conducted by Russell,
Baker and Climer of 111 owners of heat pump systems (1).
Seventeen questions were asked of owners concerning their
relative satisfaction with the heat pumps and their previous
systems. Assuming the frequency of response to particular
questions indicates which characteristics were most important,
the most important characteristics were temperature control
(36.9%), operation (28.8%), cleanliness (24.3%), and operating costs (16.2%). None of the other characteristics received as much as 10% response. It is interesting to note
that owners were more interested in temperature control,
operation, and cleanliness than in operating costs. This
suggests that researchers developing more energy-efficient
means of comfort-conditioning should pay careful attention to
the qualitative characteristics of such systems, lest their
thermodynamic or cost-minimizing masterpieces become technological curiosa rather than important contributors to
energy conservation.

The results of the heat pump survey are, of course,
suggestive rather than definitive. Much more comprehensive
research on qualitative aspects of the choice of comfort-
conditioning modes is necessary. Similar studies should be
done with surveys that represent the entire spectrum of comfort-conditioning systems -- including, for example, questions about esthetics of solar collectors and about whether
zone control of space conditioning is, or would be, used.
Also, it must be kept in mind that methodological research
on consumer choice (including qualitative characteristics
along with price, income, weather, etc.) is still in an early
stage of development and is not a straightforward econometric
modeling exercise. There are many extremely difficult questions about the construction of models of the demand for
differentiated products.

3. Social Choice Aspects of Comfort Conditioning

The primary considerations in social choice regarding
comfort conditioning involve setting the price for energy
and restricting the choice of alternative space-conditioning
systems. The motivations for social action in setting energy
prices are familiar. The earliest of these actions were
state regulation of electricity and natural gas prices to
ultimate customers and federal regulation of interstate
natural gas and electricity sales to the utilities that sell

them to ultimate customers. Federal regulation of petroleum prices is largely a post embargo phenomenon. However, for many years the United States maintained an oil import quota program designed to restrict the flow of foreign oil and keep prices for domestic producers (and therefore for consumers) artificially high.

The original motivation for regulation of the sales of electricity and natural gas was to prevent sellers of these forms of energy from exercising monopoly power. The guiding principle of such regulation is that prices reflect costs plus an adequate rate of return for the owners of the utility. In fact, however, many other factors enter into the setting of electricity and gas prices. In the case of natural gas, in particular, there is great concern about whether pricing natural gas at its production cost adequately reflects the cost of replacing it. The same type of argument is advanced in the pricing of domestic petroleum production. These concerns are exacerbated by the fact that increased reliance on imported petroleum has placed the United States in an increasingly uncomfortable balance of payments situation and an increasingly vulnerable national security position. Considerations of resource depletion, balance of payments, and national security have been used as justifications for increasing the price (and thereby discouraging use) of natural gas and petroleum products considerably beyond what extraction cost and rate-of-return considerations alone would justify.

Also, all forms of energy production and use involve, in various degrees, disruption of the natural environment and threats to public safety. Since these are costs to the public, it may be argued that the public should be compensated in the form of user or producer tax revenues for the real or potential harm imposed upon it by users or producers of energy.

How do these considerations affect the choice of comfort-conditioning systems? First, the movement toward deregulating natural gas prices and pricing domestic petroleum at the world market price implies that these scarce and versatile fuels will be priced closer to their replacement costs and to the costs of substitutes for domestic production. Therefore, the choices of space-conditioning systems will reflect choices made on grounds of economic efficiency more closely than has been the case in the past. To the extent that these price rises are more rapid than those for electricity, synthetic fuels, or other forms of energy, there will be some movement away from the use of natural gas and petroleum. This would mean less thermodynamic efficiency

in comfort-conditioning but a better overall use of scarce
resources, as reflected by their market prices. Regardless
of inter-fuel substitution, if energy prices rise relative to
other prices and relative to income, there will be an incentive to invest in equipment with higher thermodynamic efficiency and to adopt energy-conserving operating practices.

Another problem of pricing involves the electric utilities' obligation to construct generation, transmission, and
distribution capacity to satisfy the peak demands of their
customers -- even though these peak demands may be relatively
short-lived. This implies that much of the capacity is unused much of the time. As a rule, all electricity customers
pay for the peak-load capacity whether they use increased
amounts of electricity during peak periods or not. Furthermore, the "last-on, first-off" aspect of peaking capacity
implies that this capacity is usually generated by older,
thermodynamically and economically less-efficient generation
units and by special peak-load equipment, such as gas turbines fired by natural gas or petroleum products. These
turbine units are generally less thermodynamically efficient
than base-load units and are dependent on the most versatile
and expensive fuels.

There has been a great deal of discussion about whether
electricity prices should be differentiated by time of day
and season in order to impose the additional costs of peak-
load service on the peak-load users. There are many economic,
institutional, and technical barriers to be overcome before
large-scale implementation could become a reality. To the
extent that time-of-day and seasonal pricing does take hold,
there will be an incentive for peak-load users to attempt to
shift their use of electricity off the peak. Since both
winter and summer peaks are caused largely by electric heating and cooling, such differentiated rates may be expected to
have significant impact on the choice of space-conditioning
systems. One response to this problem in Europe has been
heat storage, in which a storage medium, such as some type
of ceramic material, is heating during an off-peak period.
The heat is then stored and later used during a peak period.
To the extent that heat from storage is lost to unconditioned
space, storage heating will result in some loss of thermodynamic efficiency. However, it is felt that the gains from
more efficient electricity production more than offset this
loss. Similar schemes have been advanced for cooling.

Heating and cooling devices that require backup resistance heating (solar heating devices and heat pumps, for example) present similar price problems. That is, a heat pump
will be more efficient in a thermodynamic sense than

resistance heating, but to the extent that backup heating im-
poses a peak upon the system without corresponding base-load,
load factors will go down, more peaking capacity will need to
be installed, and the price of electricity per kilowatthour
(plus the corresponding use of petroleum and natural gas)
will be likely to go up. However, one must be careful about
jumping to conclusions concerning such technologies, for even
though such peaks may be developed for heating or air conditioning, they may not coincide with the overall peak of the
electric utility system. Thus, it is possible that a morning
heating peak arising from heat pump use may tend to balance
a utility system load which peaks later in the day.

4.   General Observations about Space-Conditioning Systems

While a principal thrust of this paper has been to show
that it is difficult to make generalizations about the
effects of physical and engineering characteristics on private and social choices of space-conditioning systems, there
are some observations that seem to be worthwile. One is
that the thermodynamic characteristics of the building itself
is one of the major factors in space conditioning and may
dominate the choice of comfort-conditioning systems as far as
equipment efficiency (thermodynamic or economic) is concerned.
Insulation, weatherstripping, double glazing, and other
measures will have significant impacts on the ability of any
system to deliver efficient performance. Which of these
measures are worthwhile depends on the relative costs of
energy and of the resources required to improve the thermal
integrity of the building. Obviously, climate is the principal variable in such decisions. Measures that are cost-
effective for increasing the thermal integrity of buildings
in Minnesota would not make sense in the San Francisco Bay
Area. Also, it is clear that forced-air systems of all
kinds -- natural gas, oil, electric furnaces and heat pumps--
offer the potential of heat loss through duct work. Even in
these cases, however, one must be careful about jumping to
conclusions, for what may appear to be heat loss is not
actually a loss if the duct work involved runs through a conditioned space. Losses are most serious when ducts run beneath the floors, in attics, or along exterior walls.

Finally, decisions on equipment choice should be based
on field evidence of heating and cooling performance.
Testing under laboratory conditions often provides a dis-
torted view of what would happen in occupied buildings.
Fortunately, a good deal of such field test data is becoming
available from various sources, such as the Twin Rivers
project (2), the work of Ohio State University (3), the
Westinghouse study of heat pumps (4), and reports on other

projects. Such information should provide builders, buyers of buildings, regulatory commissions, and energy suppliers with a great deal more information about what kinds of space-conditioning systems are appropriate under what circumstances.

5. Summary and Concluding Remarks

Economic analysis of comfort-conditioning choice is complex, as this choice depends upon climate, energy prices, household incomes, building practices, and many other factors. Few purchasers of space-conditioning equipment are going to care about thermodynamic efficiency unless it is directly related to costs. As a criterion of social choice, this emphasis on cost efficiency reflects a good deal more social rationality than adhering to thermodynamic considerations as a criterion of social choice, since it recognizes the scarcity of other resources as well as of the particular fuel used to produce the comfort-conditioning services. Even so, it is wrong to assume that the purchasers of comfort-conditioning equipment have only cost in mind. It must be recognized that this is only one of many considerations in choosing a comfort-conditioning system. Other considerations, relating to comfort, ease of use, reliability, and so on, may be at least as important.

Finally, the way energy is priced is likely to have a critical bearing on the choice of space-conditioning systems. In addition to average price, the possibility of differentiating price by time of day and season is an important consideration to be reckoned with, particularly in those systems that involve electricity.

All in all, the economics of space-conditioning choice offer no easy answers, but it seems reasonable to presume that individuals and society will be better off if we begin to ask the right questions.

References

1. Howard J. Russell, Kenneth L. Baker and James Climer,
    "Terminal Interview, EPRI-AEIC Heat Pump Load Study,"
    in Paul J. Blake and William C. Gernert, Load and Use
    Characteristics of Electric Heat Pumps in Single-Family
    Residences, EPRI EA-793, Palo Alto, California: Electric
    Power Research Institute, 1978.

2. Energy and Buildings 1 (1977.78) 207-242.   The entire
    issue is devoted to this project.

3. Charles F. Sepsy, Merle F. McBride, Robert S. Blancett
    and Charles D. Jones, Fuel Utilization in Residential
    Heating and Cooling, EPRI EA-884, Palo Alto, California:
    Electric Power Research Institute, 1978.

4. Paul J. Blake and William C. Gernert, Load and Use
    Characteristics of Electric Heat Pumps in Single-Family
    Residences, EPRI EA-793, Palo Alto, California: Electric
    Power Research Institute, 1978.

COMMENTS

<u>Karl W. Boer</u> (University of Delaware): How could baseboard heating be recommended even if zone controlled ? Heat pumps should also be zone controlled (dampers). Solar collectors should not need cleaning except for a very few locations downwind from a greasy smoke source.

<u>Author's Reply</u>: There is certainly no <u>technical</u> reason that any kind of forced air system, including heat pumps, could not be controlled by dampers. The problem comes with whether damping systems can be made easy enough to use, and with sufficient precision of control at a sufficiently low price, to be an attractive alternative to the people who use the systems. Currently, control of forced air systems is normally done by a mechanical damping mechanism at the register. Casual observation suggests that few people use dampers as a means of control on a day-by-day, much less an hour-by-hour, basis. Of course, it would be technically feasible to develop a thermostatically controlled damping system, but I would imagine that the cost would be prohibitive. A complicating factor might be the effects of damping one room or another on the temperature balance in the rest of the house - especially if there is only a single thermostat governing the heat pump or other central heating system.

Solar collectors with glass or plastic covers accumulate dirt and dust as well as grease. Thus, the problem is ubiquitous; but there does not seem to be any hard information on how much degradation in performance is involved.

6

Electric Utility
Load Management

Robert G. Uhler

1.   Introduction

The economic historian Harold Underwood Faulkner noted that
civilization could be defined as "the process of configuring
nature." He added that nature reacted as much on man as man
on nature. One exaggerated example of this interaction is
the Super Dome and the spectacle it made possible in
midwinter--the Super Bowl. In that huge-climate controlled
concrete tabernacle, thousands shared their experience with
millions across continents via satellite-beamed pictures and
sound. Although the Super Bowl may not be the epitomy of
our tribal village's cultural values, the efficient comfort
conditioning of the Super Dome is a fitting metaphor for the
modern control of nature by engineers.

Thus in New Orleans, the local electric utility helped make
possible the outwitting of nature as well as the sharing of
that pseudo cultural experience. Night was turned into day,
temperature and humidity were adjusted by computer while
smoke was precipitated from the air so that twenty-two men
could play with a synthetic pig's bladder on plastic grass.
Electrical engineering and show business triumphed. Moreover, electronics--a more subtle form of electrical know-how
allowed those sitting in the warm comfort of their homes in
Green Bay, Wisconsin, to watch, perhaps with nostalgia, and
to recall earlier championship games played by real men on
frozen turfs in blinding snowstorms. For both the
spectators in the Super Dome and those at home, electricity
made it all possible and relatively enjoyable.

The economist Underwood also suggested that the "physical
environment determines to a large extent where man shall
live, what kind of work he shall do, what he may produce,
and the routes over which he must travel and transport his
products." Underwood emphasized that the physical environ-

ment, because of its influence on economic activities, determines man's social and political point of view as well as man's habits and desires. Nevertheless, man has not accepted these physical constraints passively but has alleviated hardships and lessened limitations by technical innovation--in particular the use of energy resources in both production and consumption.

Economic historians stress that much of the American continent lies between the 40 and 70 degree temperature lines--a climate supposedly conducive to producing "the most energetic and civilized races." One such historian found a close relationship between civilization and climate, particularly "between physical and mental activity and climate." Writing in the 1920's, Ellsworth Huntington claimed that temperature was the most important element, with mental activity reaching a maximum when the outside temperature averaged about 38 degrees. Moreover he said moderate temperatures were ideal, "especially a cooling of the air at frequent intervals."

Although many of these notions have been challenged, no one can deny the market penetration of air conditioning since the end of World War II. Perhaps we are a bit more mentally alert--in any case, we are certainly more comfortable and probably more productive. Thus with the help of engineering and electricity, we have progressed from sweatshops to OSHA in less than a lifetime.

Because many threads make up the tapestry of "civilization", it is inappropriate to attribute too much to a single strand. However, the role of electricity in confronting the physical aspects of nature is undeniable because of electricity's share of the total energy used by a modern society. In addition, electricity is vital to the civilization process because of its pervasiveness, ubiquitousness, flexibility in use, and its cost. Historically, the use of electricity has doubled each decade.

Two specific uses of electricity warrant attention today because they are essential to the efficient comfort conditioning of buildings: space heating and cooling. In particular, the increase in air conditioning has contributed to the expansion of many electric utility systems. Moreover these systems have become more weather sensitive. Growth in such loads contribute to the upward movement of electric rates, the problems of licensing, siting and financing costly new base-load plants, the shortage of natural gas (to run peakers) and our dependence on foreign oil. For residential comfort conditioning, electricity demands have become more "peaked" on many systems, causing load factors to deteriorate. Thus, primary energy is converted less efficiently and, in many cases, at a greater cost to consumers.

Utilities have responded to these cost pressures by examining a variety of load management techniques ranging from time-of-day pricing to the direct control of air conditioner compressors from a central location. And, also in response to customers' outcries, the nation's regulators have requested a national study of electric rate design and load controls. Similarly, President Carter has urged "rate reform." His proposals are mired in the protracted National Energy Act debate.

President Carter's national energy policy has two major considerations: the pricing of various forms of energy and the role of market prices in fostering an efficient use of resources. For electricity pricing, the then Presidential candidate said:

> Rate structures which discourage total consumption and peak power demand should be established.

> Jimmy Carter, 1976

Carter's rate design prescription was reflected in recent legislation. It can be compared to similar remarks made years ago by a respected ratemaker.

FACTORS GOVERNING RATEMAKING

FIRST: Our minimum must be not less than the cost to us; and

SECOND: Our maximum must be not greater than the worth to the consumer.

> Henry L. Doherty, 1900

Speaking for myself, and in no way representing the state regulators, utility executives, or others associated with the rate design study, I would like to offer some of my thoughts and a few general observations about the pricing of electricity.

As you may recall, the nationwide rate design study was requested by the National Association of Regulatory Utility Commissioners (NARUC) to examine electricity pricing in general and in particular, time-differentiated pricing and also direct load control techniques. In addition, the desirability of slowing growth in peak demand and the feasibility of shifting load off-peak was of particular concern and has been a major focus of our research.

2.   Issues in Pricing

The central issues in pricing can be grouped in reference to the major ratemaking alternatives under consideration in the United States (see Figure 1). These alternatives differ primarily in two dimensions:

o   Whether and to what extent costs that vary with time should be reflected in rates (e.g., in time-differentiated pricing)

o   Whether and to what extent "marginal" (or "economic") costs, as distinguished from "accounting" (or "embedded") costs, should be reflected in rates

A third dimension, limiting the mandate to serve by load controls, is also important.

Figure 1.   Rate Design Options

| RATEMAKING | COSTING | |
	ACCOUNTING	MARGINAL
Non-time Differentiated	"Fully Distributed Costs" (FDC or NTDAC)	"Long-Run Incremental Costs" (LRIC)
Time Differentiated	Summer/Winter Rates – – – – – – Time Differentiated Pricing With Accounting Costs (TDAC)	Time Differentiated Pricing With Marginal Costs (TDMC)

Load Controls

The diagram, of course, is an oversimplification but does contrast the rate design and load control options available to utility executives and regulators. The first major choice is the conceptual basis for <u>costing</u>. This has been popularized as the accounting versus marginal cost controversy with the "traditionalists" favoring accounting or fully distributed costs (FDC or NTDAC) while the "economists" prefer marginal or long-run incremental costs (LRIC).

On the <u>ratemaking</u> axis, as shown in Figure 1, the second
major choice--a time-of-use differentiation--is under
scrutiny. Here, for example, a summer/winter differential
could be accommodated with present rate forms and existing
meters. While such rates have been implemented by a number
of utility systems with pronounced seasonal peaks, a more
difficult question involves the diurnal differentiation of
rates. This would require a more sophisticated and more
expensive metering configuration. Combining the costing
alternatives with the time-of-use ratemaking choice
generates the four broad options shown in the drawing.
Other rate design refinements, such as a demand charge,
would be possible but are not depicted. Demand charges were
used for years and have a history of their own:

> Why do utilities persist in selling residential
> service on the basis of kilowatt-hours when kilowatt-
> hours are but 10 or 15 percent of the total cost of
> rendering that service? Have those grown old in the
> industry lost their grip or are the younger men
> entering it of a caliber so much softer that they dare
> not come out like men and say: "rates for service will
> consist of two parts--a readiness-to-serve charge and a
> low kilowatt-hour charge"?

> A Young Engineer, 1927
> <u>Electrical World</u>

Much more recently, an early costing proposal of the mar-
ginalist school (i.e., LRIC) was examined at great length in
Wisconsin. Still more recently, an approach that adds a
time-of-use dimension has surfaced as time-differentiated
marginal cost or TDMC. This version captures both the
marginal cost concept, as well as the notion of time-varying
costs. It is possible, however, to design time-differentiated rates based on accounting costs (i.e., TDAC).
Continuing to simplify, the quality of electric service
provided utility customers also represents a choice. This
third axis or dimension is shown in the rate design matrix
as "load controls." Customers, for example, might allow a
radio or ripple switch on their heating or cooling
appliances, particularly if the equipment has some energy
storage capability. The direct control of the switches
would be vested with the utility so that specific loads
could be centrally managed during peak periods. In
practical terms, this changes the quality of service
rendered to the customer and should be coupled with a price
incentive that represents the cost savings realized by the
utility. Other load management options, such as inter-

ruptible rates, red lights during peak periods, or similar schemes are possible.

Given this three-dimensional representation of choices, it is possible to discuss the consequences of each in practical terms. Moving from left to right, for example, raises the revenue constraint issue, as well as a problem economists call "second best." Similarly, moving downward might create needle peaks on some systems with the associated phenomena of revenue erosion and lower load factors. In addition, singling out a particular set of customers for either marginal costing or peak-load pricing poses the legal question of "undue" discrimination. And finally, the imposition of load control devices might be contrary to the mandate-to-serve requirement that underlies utility operations in many states.

Other important issues are not so easily portrayed. Many regulators, for example, hold that the provision of electric service should be reliable but at the "least" cost attainable. This, in itself, requires a difficult trade-off between quality of service, generally described as a level of reliability, and cost. Moreover, for regulated utilities, a public service commission must perform an additional balancing act. This involves the determination of a rate of return on investment sufficient to insure the continued financial integrity of the company, but does not overburden the utility's customers with excessively high electric bills. These two issues are not illustrated but are of major concern to regulators.

Three alternative approaches are most frequently proposed: (1) continue traditional non-time-differentiated ratemaking (NTDAC) practices; (2) expand the use of time-differentiated rates based on accounting costs (TDAC); or (3) adopt time-differentiated rates based on marginal costs (TDMC). Many of the issues that must be weighed are listed in Figure 2.

3.   Time-Differentiated Rates

Quite generally, time-differentiated rates appear to be a promising approach in dealing with the cyclical demand for electricity both in managing peak demand growth and peak period energy usage. In addition, direct controls seem to be a useful complement or alternative to time-of-day pricing. It should be stressed, however, that a careful cost-benefit analysis must be carried out to ensure that customers are indeed better off under a particular load management scheme.

Figure 2. Ratemaking Issues

	Ratemaking Procedure		
	Traditional	**Time-differentiated Rates**	
		Average Cost	**Marginal Cost**
Ratemaking objectives	Revenue requirements	Revenue requirements	Revenue requirements
	"Fair" cost apportionment	"Fair" cost apportionment	"Fair" cost apportionment
	Economic efficiency	Economic efficiency	Economic efficiency
	Social goals	Social goals	Social goals
	Relative emphasis on various objectives	Different emphasis from traditional	Different emphasis from average cost
Costing for total revenue requirements	Accounting or economic costs	Accounting or economic costs	Accounting or economic costs
	Test period	Test period	Test period
Costing for rate design	Company or pool	Company or pool	Company or pool
	Functionalization method	Functionalization method	Functionalization method
	Classification method	Classification method	Classification method
	Number of customer classes	Number of customer classes	Number of customer classes
	Class determination method	Class determination method	Class determination method
	Allocation method	Allocation method	Allocation method
		Allocations to rating periods	Allocations to rating periods
		Number and breadth of periods	Number and breadth of periods
		How determined	How determined
			Definition of marginal cost
			Use of LRMC or SRMC
			Planning period
			Methodology for determining LRMC
			Annualization method
Rate design	Rate forms	Rate forms	Rate forms
	Adjustment clauses	Adjustment clauses	Adjustment clauses
	Discrimination	Discrimination	Discrimination
	Consistency with other objectives	Consistency with other objectives	Consistency with other objectives and "second best" considerations
			Method for adjusting to revenue constraint
Metering and control requirements	Metering requirements	Incremental metering and control costs	Incremental metering and control costs
		Customer equipment costs	Customer equipment costs
		Transition to new rates	Transition to new rates
Customer acceptance	Customer attitudes	Customer attitudes	Customer attitudes
	Load-forecasting method	Load-forecasting method	Load-forecasting method
	Demand elasticities	Demand elasticities	Demand elasticities
		Extent of peak shifting and load reductions	Extent of peak shifting and load reductions
		Extent of revenue erosion	Extent of revenue erosion
Cost-benefit analyes	Company or pool scope	Company or pool scope	Company or pool scope
	Customer costs included	Customer costs included	Customer costs included
	Methodology	Methodology	Methodology
	Method for treating uncertainties in forecasts	Method for treating uncertainties in forecasts	Method for treating uncertainties in forecasts
	Demand and/or supply management	Demand and/or supply management	Demand and/or supply management

Admittedly, such calculations are difficult to make because of lack of data such as price sensitivity, customer acceptance, implementation costs, etc. Nevertheless, the direction is clear. Electric rates that reflect time-varying costs are desirable but should be demonstrated, reasonably, as cost-effective in a rather broad analysis that includes both the utility and its customers. For example, a high daily load factor utility may have achieved a level of efficiency that would not be increased by time-of-day pricing. This is well understood by utilities. Companies realize that off-peak rates may be appropriate at certain times but not always. For example, after a dozen years:

> Changes in the shape of the load curve brought about the withdrawal of a special off-peak rate for power service.

> Detroit Edison, 1926

4.   Marginal Costs and Efficiency

A technical point that has occupied a great deal of thought in the rate study is the role of marginal costs in determining the appropriate time-differentiated rates for electric service. Economists recognize the importance of having all major energy sources (and other factor prices) reflect, in some way, their marginal costs but certainly this is not always the case. For electricity, some desirable movement in this direction is possible on a state by state basis but, here, national energy pricing policies for all energy sources should point the way. Having federal pricing policies for oil and natural gas recognize marginal costs would complement similar efforts in electricity pricing by state regulators.

The economic theory for the rational pricing of electric energy is fairly well developed but not fully articulated. Moreover, as we take the first steps in applying such theory in actual rate cases, it is clear that we have an incomplete empirical understanding of the causality of costs and of the price elasticity of customers to various rate forms. Thus, the precise cost-effectiveness of various load management options is not certain. The added metering and control costs, however, are known and are not inconsequential.

One premise that is well grounded in economic theory is that electric rates should reflect the economic (i.e., marginal) costs of supplying that energy. In practice, however, accounting costs and a considerable amount of judgment are used in ratemaking. Closer adherence to the economic principle would encourage efficient use of primary energy, labor and capital resources. Applying the theory, however, requires compromises in order to meet other objectives of regulation, such as rate continuity, equity, and the revenue constraint. Moreover, if other energy sources diverge in price from their marginal costs, pricing electricity at its marginal costs may not be appropriate.

5.   Pricing and Controls

Given the general notion that rates should reflect costs, it follows that, to the extent that costs vary by time of day or by type of service, electric utility rates should vary by time of day and by type of service. The provision of time-differentiated rates and of different types of electric service, such as interruptible or controlled service, would serve efficiency as well as increase the choices available to consumers. In addition, it would give them an opportunity to adjust their time patterns and standards of consumption according to the real costs that are incurred in providing service. In some instances, individual customers could realize savings if offered time-of-use rates or if they tolerated shutoffs. And, as the total cost of providing service is minimized due to better use of facilities, customers in general might profit and society in a broader sense could reap economic benefits.

More particularly, load management, i.e., wider use of time-differentiated rates, load control devices, and interruptible service, should result in more efficient use of existing generating plant and should reduce the utilization of peaking generators that rely on natural gas and oil. Further, load management would permit an increase in generating plant at a rate of expansion that is consistent with optimal use of resources while still providing the amount and quality of service demanded by Americans as the economy grows. Finally, plant mix would tend towards base load generation which generally uses coal and uranium. The "old ratemaker" recognized this long ago:

> If desired, a clock arrangement can be used to throw a heavy conductor in series with the demand meter, to encourage heavy use of current at desirable hours of the day.

> Henry L. Doherty, 1900

6.   Conservation, Equity and Income

Electric rates based on economic costs also would provide
price signals that ration the use of energy thus promoting
conservation. This does not do violence to the cherished
idea of consumer sovereignty. Moreover, those imposing
costs would pay an appropriate amount for the use of
resources. And, of course, they would enjoy the comforts
and benefits of electric service in the quantity and quality
that they chose. This is both equitable and efficient.
Further, to the extent that cost-based rates reflect exter-
nalities (e.g., the costs of restoring land), environmental
concerns will be addressed.

There are, however, other issues. The impact of rising
electricity prices on economically disadvantaged people has
forced state regulators to consider the use of electric
rates for income redistribution. Lowering utility prices
below cost, however, seems to me to be an inferior and
perhaps counterproductive way of alleviating poverty. Some
wealthy individuals consume small quantities of electricity
and deserve no subsidization nor should their usage be
encouraged by low prices. Further, some low income families
consume relatively large amounts of energy and would hardly
benefit if most of their consumption were priced above
costs. Thus, while I oppose "lifeline" electric rates, I
strongly advocate welfare reform to help the needy
directly--preferably with cash payments. "Energy stamps,"
however, might be an appropriate compromise. Nevertheless,
when the difference between poverty and minimal adequate
family income is over $4000, lifeline electric rates are
hardly sufficient as an income support measure.

7.   Flexibility and Gradualism

In shaping a national energy policy, it would be disruptive
to recommend drastic changes in the structure of electric
rates. Further, it would be unduly confining to specify
exact rate designs or particular load control techniques for
all utilities. Each utility confronts a unique set of
circumstances related to local primary energy sources,
generation mix, transmission and distribution systems,
consumer requirements, load factor, growth rates, and
weather. Moreover, each regulatory commission is con-
strained by state law, procedural precedents and judicial
review. Thus, existing pricing policies reflect these
circumstances and, in turn, have influenced past investment
decisions of both utilities and customers. A gradual
movement in the right direction for ratemaking, however,
should be strongly encouraged at the Federal level.

A national energy policy must also recognize that each state
regulator has an obligation to weigh several, often con-
flicting, objectives while fashioning "just and reasonable"
electric rates as illustrated in Figure 3 below.
Regulators, in fulfilling their responsibilities, must
reconcile several major goals: providing reliable service,
minimizing costs to consumers, and ensuring fair and
equitable distribution of cost burdens among customer
classes. Further, state commissioners must preserve the
financial integrity of the utilities they regulate and must
promote economic efficiency.

Figure 3. Regulatory Objectives and Rate Design Options

REGULATORY OBJECTIVES	RATE DESIGN OPTIONS		
	Traditional (NTDAC)	Time-Differentiated (TDAC)	(TDMC)
1. Provide Reliable Service			
2. Minimize Costs to Consumers	?		
3. Fair Distribution of Cost Burdens	?		
4. Preserve Financial Integrity of Utility		?	?
5. Promote Economic Efficiency	?	?	

The reconciliation of these five goals is highly judgmental
and has a localized political dimension--commissioners are
elected or appointed at the state level. Moreover,
regulators face particularly complex problems in pricing
electricity. Costing and rate design admittedly depend on a
mixture of science and art and involve both scientific
expertise and professional judgment. These have been major
factors in rate design for decades. Consider just the
problem of allocating capacity costs to customer classes:

> One of the most difficult problems has been and
> continues to be the determination of the true cost of
> service to each group of consumers. Without a true
> knowledge of correct costs it is impossible to
> differentiate between profitable and unprofitable
> business; it is impossible to adopt the correct
> commercial policy and it is impossible to have correct
> rates.

Philadelphia Electric, 1927

There is, in fact, hardly any limit to the number
of plausible solutions that can be suggested. Equity,
from which they mostly start, will support
anything...in this field, equity is the Mother of
Confusion.

W. Arthur Lewis, 1949

Class cost of service studies would require the
selection among numerous theories, none of which are
deemed reliable. The allocation of joint costs after
they have been incurred is nearly always arbitrary.

Commonwealth Edison, 1953

There is no correct method of distributing the
costs of capital among the different classes and
quantities of service.

Clair Wilcox, 1971

8.   State Responsibilities

State commissions should continue to have primary
responsibility for the details of electricity pricing. The
federal government, however, could encourage ratemaking
principles, such as: basing rates closer to marginal costs,
using time-differentiated pricing, and offering interruptible service. In addition, it also might urge that a
cost-benefit framework for evaluating such changes in
ratemaking in particular and load management in general be
established. Further, Washington might urge broadening the
load management demonstration projects to encourage
experimentation and creativity in ratemaking and load
controlling.

Suggesting that state commissions place greater reliance on
economic principles in determining cost-based rates would
foster a more efficient allocation of resources. It should
be recalled, however, that the regulatory practice of
setting an overall revenue requirement based on accounting-
type costs will limit the extent to which marginal costs can
be reflected in rates. Nevertheless, the determination of
marginal costs is desirable because it would alert
utilities, regulators and customers to trends in costs and
in future rates and would signal clearly the cost consequences of additional consumption or the savings associated
with conservation.

9.   Judgment in Ratemaking

Two years ago, the use of marginal costs in designing
electric rates was controversial. Today, recognizing
marginal costs as a benchmark is better understood. In
fact, over the years some marginal cost pricing has been
practiced by electric utilities, and marginalism is widely
accepted in business decision-making. Thus, much of the
"great electric rate debate" has subsided as the economic
principles that underlie marginal costs are grasped by rate
engineers and the problems of application are confronted by
economists. For example:

> Attention should be given to long-run incremental
> costs in designing rates.
>
> > Edison Electric
> > Institute, 1976

> The benchmark for appropriate signals via rates is
> the social marginal cost of electrical production.
>
> > California Energy
> > Commission, 1976

> The economic theory of marginal costs is an
> analytical tool not capable of being translated into a
> ratemaking device.
>
> > National Association of
> > Manufacturers, 1976

Still, the degree to which electric rates should reflect
marginal costs in practice is debatable and, therefore, is
the type of question best reserved for detailed analysis by
state commissions. Further, these regulators are
responsible to the people most affected by such judgments.
In the state of New York, for example, the definition and
quantification of marginal cost as well as the optimum
divergence of rates from marginal cost are being examined
utility by utility in a manner that affords ample
opportunity for all of the affected parties to be heard.
This level of detail and flexibility, which is essential to
regulation in general and ratemaking in particular, is
difficult to legislate or to specify by executive order.
More generally, the New York Commission has supported the
movement to time-differentiated rates that reflect marginal
costs:

> Marginal costs do provide a reasonable basis for electric rate structures.
>
> New York Public Service
> Commission, 1976

> The LILCO cost study provides a reasonable guide to setting time-of-day rates. The company's proposal will clearly constitute important steps in the direction of greater economic efficiency.
>
> New York Public Service
> Commission, 1976

10.   The NARUC Study

Nationwide, the regulatory community has been examining these pricing and costing issues both individually by state and collectively. Many states have held formal proceedings on electric rate design either in generic hearings or in the context of particular rate cases. Moreover, several state commissions have conducted staff investigations and studies of costing methodologies and rate design. And, collectively, NARUC prompted the nationwide rate design study.

Many of the points raised here are discussed in some sixty reports prepared for the NARUC rate design study (see attached list). These have been distributed to state regulators and to federal officials, many of whom participated fully in the study itself. An overall assessment of our research was submitted in a comprehensive report to NARUC in November 1977. Additional subsidiary reports will be published over the next year.

11.   Conclusion

In sum, I believe that time-differentiated electric rates are feasible and should be gradually adopted where benefits to consumers can be demonstrated. Moreover, electric rates (and other energy prices) should reflect marginal costs to an extent consistent with other regulatory or national objectives. There are still questions to be resolved but the direction is clear: load management via price and controls will benefit all of us.

The "old ratemaker" had this to say about his approach:

> It will develop your business, repress agitation for municipal ownership and the granting of competitive franchises, meet the competition of isolated plants and

other means of illumination. I believe it will reduce
the cost of production, better your load factor, lower
the kilowatt cost to the consumer, increase the
stability of your business, strengthen your securities
and increase your earnings; in short, prove a panacea
for most of the ills to which the average utility is
heir.

GUIDE TO ELECTRIC UTILITY RATE DESIGN STUDY REPORTS*

More than fifty reports have been distributed by the Rate
Design Study. This guide is intended to help the reader
locate information more easily and thus make the reports
more accessible. During the first phase of research
(roughly 1976-77), the research was divided into ten topics
as listed below:

1. Analysis of Various Pricing Approaches
2. Considerations of Demand Elasticity for
     Electricity
3. Rate Experiments Involving Small Customers
4. Costing for Peak-Load Pricing
5. Ratemaking
6. Measuring the Potential Cost Advantages of
     Peak-Load Pricing
7. Metering
8. Technology for Utilizing Off-Peak Energy
9. Mechanical Controls
10. Customer Acceptance

For each topic, an advisory group (e.g., Task Force 2) prepared a report about a single topic. In addition, one or
more consultants worked on each topic. In some cases, a
consultant conducted research on a single topic; in other
cases, a consultant may have worked on three or four topics.
One task force (i.e., Task Force 4) produced two reports.
In a few cases, a consultant wrote more than one report on a
single topic.

Shown on the following pages are all of the reports
presently available, arranged by topic. A report order
number precedes each title.

---

*For further information, contact Rate Design Study, EPRI,
P.O. Box 10412, Palo Alto, CA, 94303.

Topic 1:   Analysis of Various Pricing Approaches

A.   Task Force 1

10.   Analysis of Various Pricing Approaches,
February 2, 1977.

B.   Consultants

9.   An Overview of Regulated Ratemaking in the United
States: Topic 1.1, February 2, 1977; National
Economic Research Associates, Inc.

7.   Analysis of Electricity Pricing in France and
Great Britain: Topic 1.2, January 25, 1977;
National Economic Research Associates, Inc.

15.   A Framework for Marginal Cost Based Time-Differen-
tiated Pricing in the United States: Topic 1.3,
February 21, 1977; National Economic Research
Associates, Inc.

14.   The Development of Various Pricing Approaches:
Topic 1.3, March 1, 1977; Ebasco Services, Inc.

Topic 2:   Considerations of Demand Elasticity for Electricity

A.   Task Force 2

12.   Elasticity of Demand,January 31, 1977.

B.   Consultants

11.   Considerations of the Price Elasticity of Demand
for Electricity, January 31, 1977; National
Economic Research Associates, Inc.

13.   Elasticity of Demand, February 10, 1977;   J. W.
Wilson & Associates, Inc.

Topic 3:   Rate Experiments Involving Small Customers

A.   Task Force 3

3.   Rate Experiments Involving Smaller Customers,
January 21, 1977.

B.   Consultant (None)

Topic 4: Costing for Peak-Load Pricing

A.   Task Force 4

22.   Comments on Two Costing Approaches for Time
      Differentiated Rates, March 8, 1977.

46.   Critical Issues in Costing Approaches for Time-
      Differentiated Rates, January 12, 1978.

B.   Consultants

23.   How to Quantify Marginal Costs, March 10, 1977;
      National Economic Research Associates, Inc.

28.   How to Quantify Marginal Costs, Results for
      Virginia Electric and Power Company, June 6, 1977;
      National Economic Research Associates, Inc.

31.   How to Quantify Marginal Costs, Results for the
      Portland General Electric Company, June 20, 1977;
      National Economic Research Associates, Inc.

35.   How to Quantify Marginal Costs, Results for the
      Dayton Power and Light Company, June 20, 1977;
      National Economic Research Associates, Inc.

47.   How to Quantify Marginal Costs, Results for
      Tennessee Valley Authority, December 16, 1977;
      National Economic Research Associates, Inc.

42.   NERA's Responses to Questions from Task Force 4,
      August 3, 1977; National Economic Research Asso-
      ciates, Inc.

58.   How to Quantify Marginal Costs: A Reply to Task
      Force 4 Comments, December 19, 1977; National
      Economic Research Associates, Inc.

24.   Costing for Peak-Load Pricing, May 4, 1977; Ebasco
      Services, Inc.

27.   Costing for Peak-Load Pricing, Results for
      Virginia Electric and Power Company, June 6, 1977;
      Ebasco Services, Inc.

30.   Costing for Peak-Load Pricing, Results for the
      Portland General Electric Company, June 20, 1977;
      Ebasco Services, Inc.

33. Costing for Peak-Load Pricing, Results for
     Carolina Power and Light Company, June 20, 1977;
     Ebasco Services, Inc.

34. Costing for Peak-Load Pricing, Results for The
     Omaha Public Power District, June 20, 1977; Ebasco
     Services, Inc.

37. Costing for Peak-Load Pricing, Results for
     Minnesota Power and Light Company, June 20, 1977;
     Ebasco Services, Inc.

41. EBASCO's Responses to Questions from Task Force 4,
     September 30, 1977; Ebasco Services, Inc.

56. EBASCO'S Responses to Major Costing Issues Raised
     by Task Force Four in its Citique of the Vepco
     Analysis, April 12, 1978; Ebasco Services
     Incorporated.

43. Comments·on National Economic Research Associates'
     Approach to Marginal Cost Pricing, September 15,
     1977; Ralph Turvey.

44. Comments on Ebasco Services' Approach to Peak-Load
     Pricing, November 28, 1977; Walter A. Morton.

Topic 5: Ratemaking

A. Task Force 5

8. Ratemaking, February 4, 1977.

B. Consultants

25. Ratemaking, June 6, 1977; National Economic
     Research Associates, Inc.

29. Ratemaking, Illustrative Rates for Virginia
     Electric and Power Company, June 6, 1977; National
     Economic Research Associates, Inc.

32. Ratemaking, Illustrative Rates for the Portland
     General Electric Company, June 20, 1977; National
     Economic Research Associates, Inc.

36. Ratemaking, Illustrative Rates for the Dayton
     Power and Light Company, June 20, 1977; National
     Economic Research Associates, Inc.

39. Making the Transition from Unit Marginal Costs to
     Rates: Results for Virginia Electric and Power
     Company, August 4, 1977; National Economic
     Research Associates, Inc.

48. Making the Transition from Unit Marginal Costs to
     Rates: Results for Portland General Electric
     Company, December 20, 1977. Prepared by National
     Economic Research Associates, Inc.

26. Ratemaking: Topic 5 and Illustrative Rates for
     Five Utilities, June 6, 1977; Ebasco Services,
     Inc.

55. Ratemaking: The Transition from Costing to Rate
     Design, April 12, 1978; Ebasco Services
     Incorporated.

Topic 6: Measuring the Potential Cost Advantages of Peak-Load Pricing

A. Task Force 6

19. Potential Cost Advantages of Load Management,
     March 4, 1977.

B. Consultants

16. Potential Cost Advantages of Peak-Load Pricing,
     February 15, 1977; Power Technologies, Inc.

59.* Potential Cost Advantages of Peak-Load Pricing,
     Volume II, Data and Computer Program Manuals,
     December 21, 1977; Power Technologies, Inc.

17. Estimating the Benefits of Peak-Load Pricing for
     Electric Utilities, February 22, 1977; Systems
     Control, Inc.

20. Demonstration of the Use of the Westinghouse Model
     LooPeak, April 15, 1977; Energy Utilization
     Systems, Inc.

21. Measuring the Potential Cost Advantages of Peak-
     Load Pricing, February 22, 1977; Gordian Asso-
     ciates, Inc.

54. Measuring the Potential Cost Advantages of Peak-
     Load Pricing (Phase B), December 15, 1977; Gordian
     Associates, Inc.

Topic 7: Metering

A.   Task Force 7

    4.   Metering, January 12, 1977.

B.   Consultant

    5.   Topic 7: Metering and Communication Systems;
Topic 8: The Utilization of Off-Peak Electricity;
Topic 9: Mechanical Controls and Penalty Pricing;
January 15, 1977; Arthur D. Little, Inc.

Topic 8: Technology for Utilizing Off-Peak Energy

A.   Task Force 8

    40.   Technology for Utilizing Off-Peak Energy,
October 15, 1977.

B.   Consultant

    5.   See Report #5 by ADL above under Topic 7.

Topic 9: Mechanical Controls

A.   Task Force 9

    6.   Mechanical Controls and Penalty Pricing,
January 14, 1977.

B.   Consultant

    5.   See Report #5 by ADL above under Topic 7.

Topic 10: Customer Acceptance

A.   Task Force 10

    2.   Customer Acceptance: Topic 10.2, January 4, 1977.

B.   Consultant

    1.   Attitudes and Opinions of Electric Utility
Customers Toward Peak-Load Conditions and Time-of-
Day Pricing. Customer Acceptance: Topic 10.1,
January 3, 1977; Elrick & Lavidge, Inc.

    38.   Attitudes and Opinions of Experimental Customers

Toward Load Management Alternatives, August 5, 1977; Elrick & Lavidge, Inc.

Other Reports:

18. Bibliography, March 21, 1977; Task Forces and the Edison Electric Institute.

49. State and Federal Regulatory Commissions' Rate Design Activities, July 12, 1977; Rate Design Study (from responses to a questionnaire sent to state regulatory agencies in December, 1975).

51. Reviewing the Rate Design Study, February 8, 1978; National Economic Research Associates, Inc.

52. Integration and Interpretation of Study Results; February 17, 1978; Ebasco Services Incorporated.

57. 1977 Survey of State and Federal Regulatory Commissions' Electric Utility Rate Design and Load Management Activities, October 25, 1977; Elrick & Lavidge, Inc.

50.* Plan of Study, September 24, 1975; Rate Design Study.

62.* Plan of Study, December 21, 1977; Rate Design Study.

Reports Submitted to NARUC:

60.* Rate Design and Load Management: A Progress Report to the National Association of Regulatory Utility Commissioners, October 28, 1976; Rate Design Study.

61. Rate Design and Load Control: Issues and Directions, A Report to the National Association of Regulatory Utility Commissioners, November, 1977; Rate Design Study.

COMMENTS

<u>Gary Kah</u> (Donovan, Hamester & Rathen, Inc.): The author
mentions the use of night-time electricity, priced only at
the energy cost, as a load management tool. Yet the recent
Office of Technology Assessment study on the On-Site
Utilization of Solar Energy describes such rates as promotional (and hence illegal in many areas) for several
reasons, including:

> such rates are the result of static analysis
> some parts of the valley are used for equipment
> maintenance
> who then pays for the day/night meter ?

Should night-time electricity be priced at its true
costs, both prorated capital charges and energy costs ? Loads
should be shifted with technology, not by demand-inducing rates.

<u>Author's Reply</u>: Notions of "true cost" are many - all rates
have an element of judgment in them. Describing some rates
as "promotional" is as empty a phrase as arguing that loads
should only be shifted by "technology" rather than "demand-
inducing rates". There is nothing wrong with using prices
that generally reflect costs to influence consumption.

<u>Jerome Strauss</u> (Versar, Inc.): Can one assume that load
management/rate design is applicable to steam distribution
by utilities for district heating systems ?   Will time-of-day
pricing be as effective with steam ?

<u>Author's Reply</u>: No comment.

7

Conventional and Solar Community Energy Systems

Henry Kelly, David Claridge, John Furber and John C. Bell

Abstract

This study examines the cost and performance of a
number of different onsite systems designed to meet the
energy requirements of a small residential community.
Systems using solar energy and systems using hydrocarbons
or solid fuels will be examined. The technologies examined
include cogeneration, district heating, direct solar heating,
solar thermal electric equipment, photovoltaic devices, and
a variety of different techniques for storing energy. A
consistent approach to costing and to economic and technical
analysis is used to provide useful comparisons between the
systems examined.

I. Introduction

It is becoming increasingly apparent that major changes
will be made in the way this nation generates and consumes
energy during the next few decades. The existing stock of
equipment, which relies heavily on plentiful and inexpensive
sources of natural gas and petroleum, will need to be
replaced, and the increase in energy prices which will force
this replacement will result in much greater attention to
the problems of energy conservation. The fact that these
changes will be made is really beyond serious question; the
only real issue is whether the transition will be made
intelligently and systematically, or whether it will be made
hastily in response to a series of crises.

Clearly, developing a strategy for this transition will
not be easy. It will require an assessment of environmental,
economic, technical, and social issues and an understanding
of constraints imposed by existing institutions, financial
realities, the stock of existing equipment, and political
possibilities. It is equally clear, however, that it would

be a mistake to be so absorbed by these constraints that
fundamental questions about which outcome would be preferred
on purely technical grounds are not asked. Analysis of
issues such as these is in a surprisingly primitive state
and many major topics remain virtually unexamined.

An analysis of energy systems for a community provides
an excellent opportunity to take a broad view of these issues.
The assessment forces considering the community as a single
integrated entity with a fixed set of energy requirements --
maintaining building interiors at desired temperatures,
providing electricity for lighting and other miscellaneous
applications, providing heat for domestic hot water and
industrial processes, etc. Systems for producing and
consuming energy should be compared on the basis of their
ability to provide energy for the same set of functions. It
is probable that any attempt to simplify the problem by
considering the capabilities of components in isolation will
result in a less effective analysis. Moreover, it is likely
that without taking this perspective, some critical aspect
of the overall system will be neglected.

The difficulty of the analysis is due in part to: the
complex and highly interdependent energy systems which have
emerged over the past few decades; the enormous variety of
equipment and fuels which are available; and, the bewildering
variety of new devices under development which may dramatical-
ly change the technical possibilities. The analysis
presented in this paper provides a quick overview of some of
the major issues confronted when examining the operation of
integrated energy systems and illustrates some of the major
design alternatives. Emphasis has been placed on reviewing
solar energy alternatives but a number of advanced fossil
systems are examined to put the solar results in perspective.
Since solar energy systems are seldom designed to provide all
energy requirements of a building or community, it was also
necessary to examine how the solar devices could best be
integrated into conventional systems of energy supply.

II. System Design Requirements

Several very basic choices must be made when designing
an energy system:

Generation

The optimum size and location of generating equipment
used must be determined. Possible alternatives include
constructing conventional centralized electric generating
facilities, synthetic fuel facilities, large solar generating

plants, or small onsite systems. The choice is particularly difficult in the case of solar energy since the solar resource is available equally throughout the community.

Transmission

If energy is transmitted, it is necessary to determine whether it should be transferred in thermal, electrical, or chemical form or whether several different transport systems should be employed simultaneously. Again, the choice is difficult for solar facilities since solar devices can provide energy in each of these forms.

It is apparent that the efficiency of energy utilization in the community can be improved by connecting as many customers and generating facilities together as possible. Such connections increase the control which can be exercised over the system. The load leveling which occurs when demands are combined can increase the load factor for the electric generators used in the network. Larger load factors reduce the peak generating capacity required per unit of energy produced and thereby reduce the effective capital cost of generating devices. Improved load factors also reduce inefficiencies attributable to part-load operation. It is important to notice, however, that these effects apply to small as well as to large generating devices integrated into the system.

The cost of energy transport is frequently ignored, but it is an important consideration which mitigates the improved performance resulting from interconnection of the systems. Over 2/3 of the present electric energy capital costs are attributable to transmission and distribution equipment, and while this ratio is decreasing as generating equipment becomes more expensive, the cost and performance of energy transport is clearly too important to ignore. Comparing the real cost of different energy transportation systems is difficult because costs are extremely sensitive to local conditions. Trenching, for example, can be ten times more expensive in an urban area than it is in rural regions. Table 1 illustrates some of the factors which must be considered. It is obvious that transporting energy in chemical form (natural gas, for example) is by far the least expensive option. Distributing energy in the form of hot water over distances of 1-2 miles, however, may be only about 30% more expensive than transmitting electrical energy over typical distances from generating facilities to consumers. In this comparison no attempt was made to share the cost of the trench dug for the hot water pipe with potable water, sewer, telephone or other lines which could be

Figure 1.   Possible Locations for Storage Equipment in Onsite Solar Energy Systems

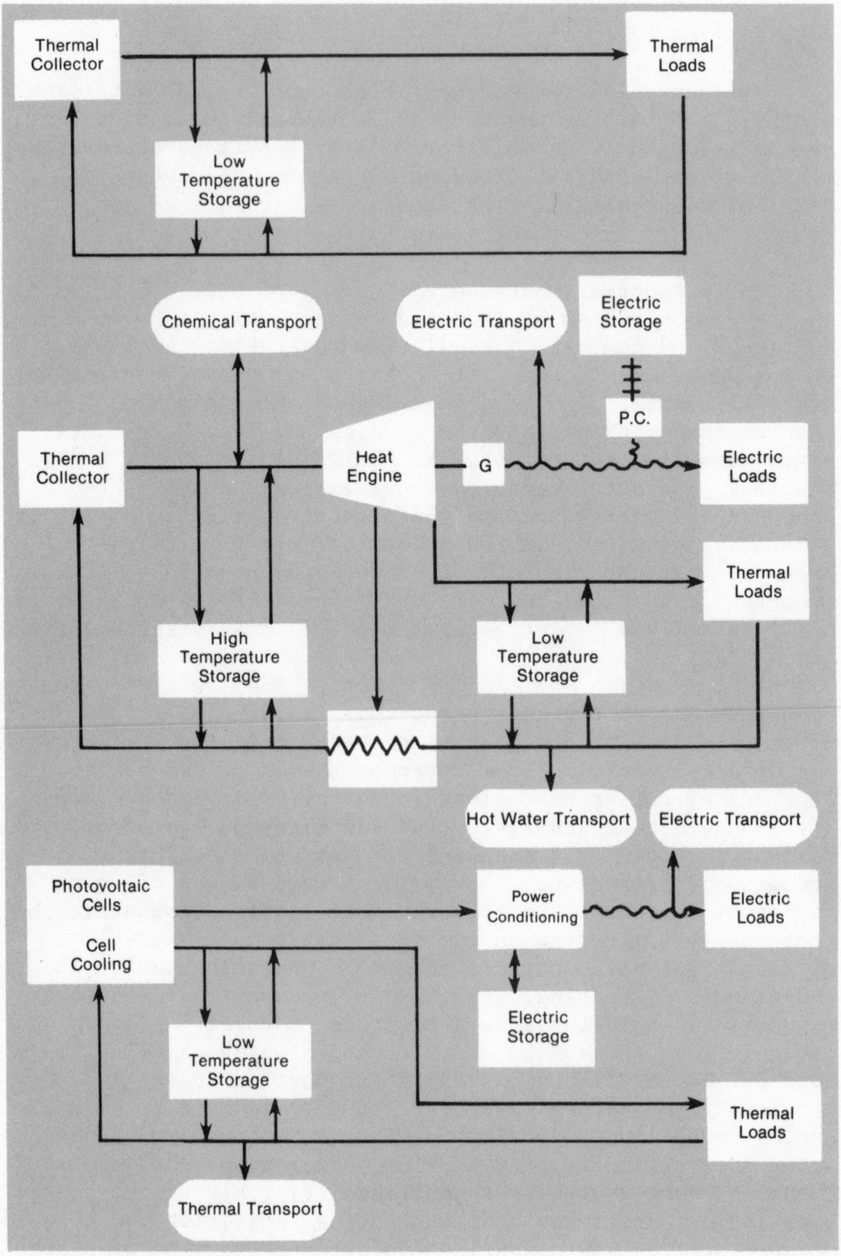

placed in the excavation. The electric distribution costs
would have been significantly higher if buried cables were
used.

Storage

Determining the best way to use storage is one of the
most difficult, and least understood, of the major issues
confronted in designing an integrated energy system. Storage
can serve a variety of purposes. The most obvious is to use
it as a buffer between the energy supply system and a varying
demand for energy. The increase in generating load factor
again reduces capital costs for the generating equipment and
improves its performance. Storage can also improve the
performance and reduce the peak capacity demand of energy
consuming equipment. Operating air-conditioners at night
and storing chilled water can, for example, significantly
improve the efficiency of the chilling devices as well as
shift electrical demand for cooling to periods when the
demands placed on the electric utility are relatively low.
In solar equipment, of course, storage can also even out the
fluctuating output of devices operating in uneven sunlight
and provide energy when the sun is not shining.

It must be recognized that most of the comments made
about the benefits of improving load factors and reducing
peak demands apply primarily to electric generating systems
since thermal and chemical energy can already be stored
relatively inexpensively.

Determination of the proper size, type, location, and
strategy of operation for storage systems integrated into a
network of energy consumers and suppliers is a problem in
applied mathematics which has not received adequate attention.
The problem is complex even when dealing with a single user
since three types of storage can be used in a system as
shown in the generalized schematic in Fig. 1. Extending
the problem to a group of users increases the complexity since
small local storage devices can be charged from a central
generating facility or vice versa.

Operational Control

A final major issue is the strategy for controlling and
operating the equipment available in the community system.
Conventional electric utilities take some care to schedule
the operation of plants to minimize costs but the problem
becomes more complex as the variety of generating and
storage devices available in the system increases. The

Table 1. A Comparison of the Cost of Transmitting and Distributing Energy in Electrical, Chemical, and Thermal Form

Mode	Capacity	Capital cost	Efficiency	Load factor[1]	Operation and maintenance	Usable energy cost[2]
Electric transmission (765kV, 500 miles)[3]	8.1×10^6 kW (3.25 lines equivalent)	$92/kW[4]	0.95[5]	0.7	5.7×10^{-4}/kWh[6]	5.2×10^{-3}/kWh[7,2]
Natural gas transmission (30 inches diameter, 800 PSI, 500 miles)[8]	$8.1 + 10^6$ kW (1 line)	$17.6/kW[9]	0.98[10]	0.88[11]	3.8×10^{-4}/kWh[12]	9.1×10^{-4}/kWh[13]
Electric distribution (10,000 customers—residential/commercial)	5×10^4 kW (17.2×10^3 kWh per customer)[14]	$130/kW[15]	0.94[5]	0.4	9.3×10^{-4}/kWh[16]	9.0×10^{-3}/kWh[7] (Total electricity = 0.0142)
Natural gas distribution (10,000 customers—residential/commercial)	8.3×10^4 kW (121 Mcf per customer)[17]	$50/kW[17]	0.98[10]	0.5	3.8×10^{-4}/kWh[18]	2.0×10^{-3}/kWh[13] (Total gas = 0.00291)
Hot water distribution (10,000 customers—residential/commercial)	7×10^4 kW[19] (21.3×10^3 kWh per customer)	$260/kW[19,20]	0.85[19]	0.35	2.6×10^{-3}/kWh[21]	1.8×10^{-2}/kWh[7]

NOTES:

[1] A capital recovery factor of 0.15 is used to calculate annual capital charges.

[2] Assumes a capacity of 2,500 MW for one 765 kv line.

[3] Utility construction expenditures of $1.7 billion in 1975 for 3,762 additional miles or $461,000 per mile average. (*Statistical Yearbook of the Electric Utility Industry for 1975*, Edison Electric Institute, New York, N.Y., Oct. 1975.)

[4] *The 1970 National Power Survey*, Federal Power Commission, Washington, D.C., p. 1-13-8, Dec. 1971.

[5] Investor-owned electric utilities spent approximately $850 million on transmission O & M costs in 1975 for 1.5×10^{12} kWh. (*Statistical Yearbook of the Electric Utility Industry for 1975*).

[6] Assumes an end-use efficiency of 100 percent.

[7] *National Gas Survey*, U.S. Federal Power Commission, Vol. 1, p. 34, 1975.

[8] All natural gas companies spent $531 million in 1976 for 1,845 miles of new transmission pipeline or $287,000 per mile average. (*1976 Gas Facts*, American Gas Association, Arlington, Va., 1977).

[9] Four percent of total natural gas consumed was used for pipeline fuel in 1976. This is equally allocated to transmission and distribution. (*AGA Gas Facts*).

[10] *National Gas Survey*, U.S. Federal Power Commission, Vol. III, p. 129, 1973.

[11] All natural gas utilities spent $1.1 billion in 1976 on O & M for transmission for 14.8 trillion cubic feet (TCF).

[12] Assumes end-use efficiency of 65 percent.

[13] "Residential Energy Use to the Year 2000: Conservation and Economics," Oak Ridge National Lab, Report ORNL/CON-13, Oak Ridge, Tenn., Sept. 1977.

[14] Investor-owned electric utilities spent $2.8 billion on construction of distribution facilities for 21,700 kW of new capacity in 1975. (*Statistical Yearbook of the Electrical Utility Industry for 1975*.)

[15] Investor-owned utilities spent approximately $1.59 billion in 1975 for distribution O & M costs for 1.5×10^{12} kWh. (*Statistical Yearbook of the Electrical Utility Industry for 1975*.)

[16] Calculated from the average cost of $400 per customer (private communication—American Gas Association) with an average hot water and space heating requirement of 36.4×10^3 kWh (121 MCF) per year and an assumed load factor of 0.4.

[17] All natural gas utilities spent $1.1 billion on distribution O & M in 1976 for 14.8 TCF (*AGA Gas Facts*).

[18] See volume II.

[19] "Evaluation of the Feasibility for Widespread Introduction of Coal into the Residential and Commercial Sectors," Exxon Research and Engineering Co., Linden, N.J., Vol. II, p. 6-11, Apr. 1977. These two studies give a construction cost of about $14 million for the size system in question, which requires a peak capacity of 70,000 kW, as shown by reference in note 20.

[20] Annual O & M costs are calculated by assuming they are 3 percent of capital costs, which is the average of the percentages for gas and electric systems.

[21] The cost of energy lost in transmission was estimated using 0.04c/kWh for electricity and 1.5c/kWh for thermal energy.

problem becomes even more complex if an attempt is made to
alter the pattern of energy demand as well as patterns of
energy supply. The performance of the integrated system can
be improved if customers postpone discretionary consumption
to periods when energy demands are relatively low. Some of
these changes will not require any change in the style of
energy consumption. Large refrigeration devices and water
heaters, for example, can be designed to operate effectively
if electricity is not available for their operation during
predictable periods during the day. Voluntary changes in
patterns of consumption can also be expected if energy costs
are higher during periods of peak electric demand. The strategy of shifting demand, of course, depends critically on an
analysis of the operation of the integrated energy system.
If solar energy is widely used in the community, for example,
it would probably be desirable to shift loads to periods of
maximum sunlight while in most other systems it would
probably be preferable to shift discretionary demand to the
late evening or early morning.

III. System Assumptions

Having laid out some of the major design choices, the
remainder of this paper will be devoted to a discussion of
specific systems chosen to illustrate these themes. As the
analysis compared the performance of a large number of
different approaches, it was not possible to perform a
detailed optimization for any of the systems assessed.

Comparing the life-cycle cost of systems using solar
energy with those of conventional systems requires knowledge
of the future price of conventional fuels. Estimates of the
future price of oil, gas, and electricity vary greatly
because of uncertainties such as the rate of future oil
discoveries here and abroad, the stringency of environmental
controls, and political decisions made by international
energy suppliers. Given the uncertainties about such a
critical variable, it was necessary to compare costs using
several different forecasts.

Energy Price Projections

The three energy price forecasts used in most of the
comparisons in this paper are summarized in Table 2. The
different assumptions for each projection are:

-- Projection (1): The cost of electricity and fossil
fuels will increase at the pace of general inflation (5.5
percent in this analysis).

Table 2. Typical Assumed Conventional Energy Costs in the
Year 2000 (in 1976 dollars)

	Projection 1* (No Escalation)	Projection 2	Projection 3
Electricity**-$/kWh$_e$			
Albuquerque-House	.024	.035	.080
-Community	.027	.039	.088
Boston-House	.044	.064	.144
Ft. Worth-House	.027	.039	.088
Omaha-House	.025	.036	.082
-Community	.023	.033	.075
Gas-Small Users $/kWh$_t$			
Albuquerque	.0050	.011	.016
Omaha	.0037	.0082	.012
Fuel Oil-Large Users $/kWh$_t$			
Albuquerque	.0063	.0089	.0207
Omaha	.0055	.0080	.0179

* Actual 1976 rates

** These average values are representative of the more
elaborate rate structure actually used in the computer model.
The model used actual utility rates for the region including
demand charges and declining block rates when these features
applied in the region.

House = prices charged for single family house using a
        heat pump.
Community = average price for all buildings in the heat
        pump community.

Table 3. Levelized Monthly Energy Costs for Single Family House in Albuquerque with Solar Heating and Hot Water

	Percent Energy Saved	ENERGY PRICE PROJECTION #1	#2	#2 20% ITC on solar equipment	#3 20% ITC on solar equipment
1) All Electric Single Family House with Heat Pump	0	$157	$204	$204	$395
2) Standard Solar Heating and Hot Water System (45 m² of collectors) Heat Pump Backup					
--High Cost	48	188(237)	215(263)	202(253)	310(361)
--Low Cost	48	159(195)	185(221)	177(214)	285(322)
3) Solar Heating and Hot Water with Central Seasonal Storage 40 m² Collector per House					
--High Cost	65	216(286)	236(305)	213(286)	292(364)
--Low Cost	65	166(211)	185(231)	172(219)	251(298)

--Houses have electric hot water heaters and heat pumps except (3) which only has electric chiller.
--Parentheses ( ) indicate private utility ownership of solar equipment.
--ITC = Investment Tax Credit.

-- Projection (2):   Energy prices will rise at rates
projected by Brookhaven National Laboratory (BNL)(1).
Electricity prices are expected to rise by about 41 percent
(in constant dollars) by the year 2000 (to roughly the
current marginal cost of electricity from new plants) and
gas prices to increase by 123 percent during the same
interval.

-- Projection (3):   Energy prices will increase by a
factor of 3.4 by the year 2000. Under this assumption, the
price of oil and gas in most cities would be roughly equal to
the price of synthetic fuels. Electricity rates would
increase to $0.07 to $0.10/kWh in all cities examined except
Boston, where the price would be somewhat higher.

Investment Tax Credit

Investment tax credits have received considerable
attention as a method for lowering the cost of solar energy.
About half of the comparisons made assume that investments
made in solar equipment or energy conservation equipment
receive a 20% tax credit.

IV. Results of the Analysis for Single Family Houses

Table 3   (and most of the tables which follow) shows
two types of variables. The first column indicates the
percentage of the total primary energy demand of the building
supplied from solar energy sources if any solar energy is
used. It should be noted that the number shown is the ratio
of the total useful solar energy to the total energy consumed
by the building for all purposes -- heating, cooling, hot
water, lighting, etc. Therefore, a system providing 100% of
all heat and hot water in Albuquerque, provides only 65%
of the total energy needs of the building. The remaining
columns indicate the levelized monthly cost of energy
provided to meet the end-use demands of the house for
different energy price/tax credit combinations. This cost
includes capital charges, the cost of replacing major
components which wear out, operating costs, and charges for
fuel and electricity purchased. A "levelized monthly cost"
is defined to be the sum which, if paid in equal amounts for
thirty years, would have the same present value, discounted
at the consumer's discount rate, as the actual cash flow
experienced by the consumer over this period. The economic
methodology is presented in more detail in Appendix A.
The numbers in parentheses indicate the price the customer
would pay if a privately owned utility financed the solar
equipment and included its cost of capital in the customer's
energy bill.

Energy Transportation and Storage

Table 3 examines the twin issues of energy transport and energy storage. In this system two sources of energy are used: flat plate collectors installed on the roofs of individual residences, and a conventional central electric utility consisting of a mixture of coal, nuclear, and oil fueled generating plants. (Detailed assumptions used to perform the analysis shown here are available in the OTA report, Application of Solar Technology to Today's Energy Needs.)(2)

The first system shown in the table indicates the costs experienced in a house equipped with a conventional electric heat pump and a standard electric water heater. The second system is a standard solar heating and hot water system using 45 m^2 of flat plate collectors and a concrete hot water storage tank in the basement. This system can provide 48% of the total energy requirements of the house at a cost which is competitive on a life-cycle basis if it is assumed that the price of electricity increases modestly. (The range of solar prices reflects different assumptions about the cost of collectors and other solar components.) The third system examined reflects the costs of a system consisting of 300 identical houses connected to a central hot water storage facility with a piping system. A central storage facility was used because the larger storage facility is significantly less expensive per unit of energy stored. Economies of scale tend to diminish, however, as storage facilities become much larger than the device used in the 300 house system. Piping costs per unit also tend to increase as the number of buildings connected increases. (This study made no attempt to determine the optimum number of buildings for such a system.) The equipment was able to provide 100% of the heating and hot water requirements of the 300 house cluster at approximately the same cost as the solar houses connected only with electrical distribution systems. It is interesting to notice that the collector area required for the 100% solar heating and hot water system is actually smaller than the collectors used for the individual solar houses which provided only 74% of the heating and hot water (48% of total energy).

Conventional Backup Costs

Table 4 explores another set of issues confronted in connecting the small, onsite generating units with centralized electric utilities. In this case the real cost of providing backup electric power to each building type was estimated by computing the incremental costs borne by a utility if

Table 4. Levelized Monthly Costs for a Single Family House Showing the Effect of Marginal Costing and Buying of Offpeak Power

	Electricity rates assumed to increase by BNL forecast		Electricity rates reflect marginal utility rates (see appendix for methodology)	
	No credits	20% ITC on solar & offpeak equipment	No credits	20% ITC on solar & offpeak equipment
I. Albuquerque				
—No solar or offpeak buying:				
• Electric resistance heat....	239	239	241	241
• Heat pump heat	204	204	229	229
—Solar only:				
• Low cost colls	185(211)	179(205)	183(209)	177(203
• High cost colls	205(240)	196(232)	203(238)	194(230)
—Offpeak heating, & hot water only	N/A	N/A	185(205)	181(201)
—Offpeak heating, cooling & hot water only..............	N/A	N/A	198(232)	189(225)
—Solar & offpeak heating & hot water				
• Low cost colls	N/A	N/A	177(207)	169(200)
• High cost colls	N/A	N/A	197(236)	186(227)
—Solar & offpeak heating, cooling & hot water				
• Low cost colls	N/A	N/A	194(242)	181(231)
• High cost colls	N/A	N/A	214(272)	198(258)
II. Omaha				
—No solar or offpeak buying:				
• Electric resistance heat....	278	278	278	278
• Heat pump heat	250	250	268	268
—Solar only				
• Low cost colls	252(282)	244(276)	234(265)	227(259)
• High cost colls	278(320)	267(311)	261(303)	250(294)
—Offpeak heating & hot water only	N/A	N/A	219(241)	214(237)
—Offpeak heating, cooling & hot water only..............	N/A	N/A	237(278)	227(270)
—Solar & offpeak heating & hot water				
• Low cost colls	N/A	N/A	221(258)	211(250)
• High cost colls	N/A	N/A	247(296)	234(284)
—Solar & offpeak heating cooling & hot water				
• Low cost colls	N/A	N/A	244(303)	228(289)
• High cost colls	N/A	N/A	270(341)	250(324)

Solar systems are thermal only

1000 identical houses equipped with the systems indicated were added to the utility grid and dividing this cost by the total number of kilowatt hours consumed by these buildings. In each case it was assumed that the utility was able to optimize the mixture of generating equipment installed to meet the new demand. (It must be emphasized that the techniques used to construct estimates of incremental costs were approximations of the techniques actually used by utilities to determine the optimum mix of generating facilities; each utility will need to make independent estimates of costs since these effective marginal costs are sensitive to local conditions.)

Offpeak Purchase of Electricity

Table 4 can be used to understand the real costs of solar heating systems integrated with an electric utility when low-cost thermal storage is available. Four basic systems are shown: (1) a conventional system using electric resistance heating, electric hot water heating, and window air-conditioners; (2) an identical system which uses a device for storing electricity purchased at night (offpeak power) for building heating and hot water during the day; (3) a system which purchases offpeak power for most of the space conditioning and hot water demands; and (4) a solar heating and hot water system equipped to purchase backup power for heating and hot water only at night. (It is a relatively simple matter to convert a solar system to purchase power during offpeak periods since thermal storage is already available in most solar heating systems). It should be noted that with contemporary equipment, offpeak thermal storage cannot be charged with an electric heat pump because the heat pump heating cycle becomes very inefficient when the heat pump is used to achieve temperatures high enough to make use of inexpensive thermal storage. It would, of course, be possible to use resistance heating elements in the heat pumps to charge storage but this variant was not examined quantitatively. It is clear from the table that the solar system is only attractive when compared to the systems which purchase offpeak power if a low-cost solar collector is available. It must be remembered that the "marginal cost" analysis shown here assumes that utilities are able to make precise estimates of the cost of providing service to each kind of customer and no attempt was made to estimate the additional cost of metering and billing which would be incurred if the marginal charges were actually used to compute electric rates. In practice, therefore, none of the customers would enjoy as great an advantage from offpeak rates as those shown in the table.

Natural Gas Backup

Table 5 compares monthly costs of systems with electric backup with the costs of systems provided with gas backup. It can be seen that the gas systems are attractive even if the price of gas increases much more rapidly than the price of electricity as assumed by Projection (2).

Storage of Electricity

Table 6 examines a problem very similar to the one confronted in the case of solar heating systems. In this case, however, the issue is whether a small photovoltaic generating device should attempt to store electricity onsite and whether it should be allowed to sell energy to an electric utility. Cost comparisons are provided for several different assumptions about increases in electricity cost from conventional systems and for the assumption that "marginal" electric rates are charged. The table indicates that it is not desirable to store electricity in small battery systems even if battery prices are reduced to about 60 $/kwh as assumed in the calculation. This conclusion might be reversed, however, if very low-cost electricity storage becomes available because of successful development of an iron-redox battery or other advanced battery systems. The poor performance of the battery is attributable in part to the fact that electric storage is typically only about 75% efficient and in part to the fact that photovoltaic electricity tends to be available during periods of high electricity demand and is therefore relatively valuable. Sales to the utility appear preferable to onsite storage in the case examined even though under the assumptions of this analysis the utility pays only 50% as much for the electricity it purchases as it charges for the electricity it sells.

Results of Analysis for Community Systems

The final set of tables examines a series of systems designed to provide all the energy needs of a residential community with a population of 30,000. The results of the analysis are somewhat more difficult to interpret than the cases examined previously since many of the interactions explored for individual buildings occur simultaneously.

Community Assumptions

The community examined is assumed to be roughly square and about 2 km on a side. The number of buildings of various types contained in the community are summarized in

Table 5. Levelized Monthly Costs of Several Kinds of Energy Equipment in a Single Family Detached Residence in Albuquerque, N.M.

| | | ENERGY | PRICE | PROJECTION | |
	Percent Energy Saved	#1	#2	#2 20% ITC on solar equipment	#3 20% ITC on solar equipment
1) heat pump electric h.w.	0	$157	$204	$204	$395
2) solar heat and h.w. heat pump backup					
--high	48	188(237)	215(263)	202(253)	310(361)
--low	48	159(195)	185(221)	177(214)	285(322)
3) extra insulation 59 m^2 photovoltaics @ 500 $/kw, electric heat pump backup					
--no batteries	52	171(213)	197(240)	187(231)	295(389)
--20 kwh batteries @ 70 $/kwh	45	192(243)	216(267)	204(257)	304(357)
4) gas heat and h.w. central electric air conditioning	0	177	174	174	287
5) solar heat and h.w. gas backup, electric air conditioning					
--high	41	173(222)	202(251)	190(240)	277(328)
--low	41	144(179)	173(208)	165(201)	252(289)
6) extra insulation 59 m^2 photovoltaics @ 500 $/kw, gas fired heat pump/generator backup	--	125(168)	145(188)	134(179)	152(197)

All costs in 1976 dollars; ( ) = utility ownership; ITC = Investment Tax Credit to solar equipment

Table 6. Levelized Monthly Energy Costs for a Well-Insulated Single Family House Showing the Effect of Marginal Costing for Backup Power

	ELECTRICITY RATES FOLLOW PROJECTION # 2		ELECTRICITY RATES REFLECT MARGINAL UTILITY RATES	
	No ITC	20% ITC on Solar Equipment	No ITC	20% ITC on Solar Equipment
I. ALBUQUERQUE				
--No solar	$183	$183	$204	$204
--59 m² silicon photovoltaics, no batteries, no sales to the utility	213(255)	202(246)	214(257)	204(248)
--59 m² silicon photovoltaics, no batteries, sales to utility permitted	197(241)	187(232)	190(232)	179(223)
--59 m² silicon photovoltaics, 20 kwh batteries, sales to utility permitted	215(267)	202(257)	194(245)	181(234)
II. BOSTON				
--No solar	300	300	215	215
--59 m² silicon photovoltaics, no batteries, no sales to the utility	319(362)	308(353)	215(259)	204(249)
--59 m² silicon photovoltaics, no batteries, sales to utility permitted	304(348)	293(340)	200(244)	189(234)
--59 m² silicon photovoltaics, 20 kwh batteries, sales to utility permitted	323(376)	310(366)	227(279)	213(267)

Table 6 (continued)

	ELECTRICITY RATES FOLLOW PROJECTION # 2		ELECTRICITY RATES REFLECT MARGINAL UTILITY RATES	
	No ITC	20% ITC on Solar Equipment	No ITC	20% ITC on Solar Equipment
III. FT. WORTH				
--No solar	203	203	209	209
--59 m² silicon photovoltaics, no batteries, no sales to the utility	228(272)	216(262)	227(272)	216(262)
--59 m² silicon photovoltaics, no batteries, sales to utility permitted	217(263)	207(254)	211(257)	201(248)
--59 m² silicon photovoltaics, 20 kwh batteries, sales to utility permitted	238(293)	226(282)	240(295)	228(284)
IV. OMAHA				
--No solar	208	208	219	219
--59 m² silicon photovoltaics, no batteries, no sales to the utility	241(284)	230(275)	238(282)	228(273)
--59 m² silicon photovoltaics, no batteries, sales to utility permitted	231(276)	221(267)	223(267)	212(258)
--59 m² silicon photovoltaics, 20 kwh batteries, sales to utility permitted	251(305)	239(294)	250(303)	237(291)

-- All houses use heat pumps for space conditioning and electric resistance hot water heating.

-- Parentheses ( ) indicate utility ownership.

-- ITC = Investment Tax Credit.

Table 7. The table also indicates that about 0.5 km^2 of area is available for solar collectors on southern facing roofs and parking facilities if the community has been carefully designed for the use of solar energy. Another 0.25 km^2 could be available if all roads were covered with collectors.

As in the previous cases, the different systems will be compared on the basis of the charges made to the energy consumers in the community. Three "conventional" communities were selected for reference: (1) a community with a mixture of heating and cooling systems roughly in proportion to the

Table 7. Buildings in the Community of 30,000

	NUMBER OF BUILDINGS	TYPICAL AREA AVAILABLE ON SOUTHERN ROOF (m^2)	AREA AVAILABLE ON ROOF AND PARKING LOT
Single Family Detached Houses	1,864	81,600	81,600
8-Unit Town Houses	232	75,800	150,000
36-Unit Low Rise Apartments	72	48,500	103,400
196-Unit High Rise Apartments	20	20,000	103,000
Shopping Center	1	28,800	63,000
TOTALS	2,189	254,700	501,000

Ground Area Required for 100 Percent
Solar System in Albuquerque

	Area Needed (m^2)
Parabolic Dishes/Stirling Engines	800,000–1,000,000
Photovoltaic Concentrator System (2-Axis Tracking)	1,400,000–1,800,000
Pond Collectors/ORCS Engine	1,900,000–2,500,000

mixture actually occurring in the area being examined; (2) a
community in which all buildings are assumed to use electric
resistance heating and electric air conditioning; and (3) a
community in which all single family houses, town houses, and
low-rise apartments use heat pumps, and other buildings use
electric resistance heating.

V.   Results of the Analysis for Communities

The results of the analysis are summarized in Tables 8 and
9. These tables provide estimates of the levelized monthly
costs perceived by residents in the community, and an estimate
of the effective cost of solar energy (in ¢/kWh). The effective
cost is computed by dividing the difference between the annualized equipment costs for the solar and conventional systems
by the total energy saved by the solar system. The "reference
system" used in this comparison is a community identical in all
respects to the solar community but served by a conventional
electric utility. Extensive use of heat pumps is assumed in
the "reference community." (Details of the techniques used in
the calculations are provided in Appendix B.)

The specific systems chosen for examination are illustrated
in Tables 10-17. These tables provide an itemized cost list,
a system diagram, and a map of the community. Appendix C
contains additional information useful in the interpretation of
these tables.

The first systems compared to the reference system in
Tables 8 and 9 are total energy systems using conventional fuels
(oil, gas, and coal). These systems are designed to provide
100% of the energy requirements of the communities. They
generate electricity and use the heat extracted from the diesel
exhaust and cooling jacket to provide heating and cooling
(through absorption chillers) to individual buildings through
four-pipe distribution systems. The diesel systems shown are
equipped with organic Rankine cycle bottoming engines which
generate additional electricity with the heat discarded by the
diesel engine systems when the demand for direct use of this
heat is small. In these cases the "effective cost" of energy
represents the additional cost of installing the total energy
system divided by the amount of energy saved by the system.
One of the diesel systems is illustrated in Table 10.

A small map of the community examined is also provided.
The system shown in Table 2 uses a standard Rankine cycle steam
turbine with a coal boiler. Heat is extracted as needed to
operate the district heating system.

Table 8. Summary Description of System: Albuquerque Community

Summary description of system	Table number	Percent solar	Effective cost of solar energy (¢/kWh) No credits	Effective cost of solar energy (¢/kWh) 20% ITC	Levelized monthly cost of energy service Projection 1 No credits	Levelized monthly cost of energy service Projection 2 No credits	Levelized monthly cost of energy service Projection 2 20%ITC	Levelized monthly cost of energy service Projection 3 20%ITC
ALBUQUERQUE COMMUNITY								
Conventional systems								
Conv. Heating and Cooling Systems in Each Building; Mixture of Gas/Electric Hot Water, Gas/Heat-Pump/Resistance Heating, and Electric Chilling		0.0	NA/ NA	NA	90./ NA	126./ NA	126./ NA	225./ NA
Conv. Heating and Cooling Systems in Each Building, All Use Electric Hot Water, Resistance Heating, Electric Cooling, and Utility Electricity		0.0	NA/ NA	NA	129./ NA	174./ NA	174./ NA	357./ NA
Reference system								
Conv. Heating and Cooling Systems in Each Building; All Use Electric Hot Water and Cooling; the High Rises and Shopping Center Use Resistance Heating, Other Buildings Use Heat Pumps		0.0	NA/ NA	NA	125./ NA	164./ NA	164./ NA	325./ NA
Systems compared to reference system								
Conv. Engine Cogeneration; Oil-Burning Diesel/ORCS, Absorption and Electric Chillers	10	54.0	5.33/ 8.08	4.68/ 7.31	122./ 152.	135./ 166.	128./ 157.	189./ 218.
Conv. Engine Cogeneration; Gas-Burning Diesel/ORCS, Absorption and Electric Chillers		54.0	5.33/ 8.08	4.68/ 7.31	101./ 131.	118./ 148.	111./ 139.	126./ 155.
Conv. Engine Cogeneration; Coal Steam Turbines and Absorption and Electric Chillers.	11	41.7	9.18/ 13.85	8.02/ 12.45	125./ 165.	136./ 175.	126./ 164.	157./ 195.
100-Percent Solar Heating; 1-Cover Pond, Seasonal Aquifer Storage, Electric Chillers, and Utility Electricity.	12	54.7	6.67/ 9.86	5.90/ 8.94	140./ 175.	155./ 191.	147./ 181.	210./ 244.

Table 8 (continued)

Summary description of system	Table number	Percent solar	Effective cost of solar energy (¢/kWh)		Levelized monthly cost of energy service			
			No credits	20% ITC	Projection 1 No credits	Projection 2 No credits	Projection 2 20% ITC	Projection 3 20% ITC
100-Percent Solar Heating; 1-Cover Pond, Seasonal Aquifer Storage, Electric Chillers, and Utility Electricity.		54.7	5.52/ 8.28	4.88/ 7.50	127./ 158.	143./ 173.	135./ 165.	199./ 228.
100-Percent Solar Heating; Flat-Plates (1977 Prices), Seasonal Low-Temp. Storage, Absorption Chillers, and Utility Electricity.		67.0	7.82/ 11.90	6.81/ 10.68	157./ 213.	166./ 222.	153./ 205.	191./ 244.
100-Percent Solar Heating; Flat-Plates (Future Price), Absorption Chillers, Low-Temp. Storage, and Utility Electricity.	13	66.9	5.71/ 8.90	4.93/ 7.96	128./ 172.	138./ 181.	127./ 168.	165./ 207.
Solar Engine Cogeneration; Steam Turbines, Heliostats, High- and Low-Temp. Storage, Absorption and Electric Chillers and Coal Backup.	14	70.1	7.91/ 11.56	6.98/ 10.44	150./ 203.	156./ 208.	143./ 192.	158./ 208.
Solar Engine Cogeneration; Heliostats, Steam Turbines With Coal Superheat, High- and Low-Temp. Storage, Absorption and Electric Chillers, and Coal Backup.		66.4	7.75/ 11.37	6.83/ 10.26	144./ 193.	150./ 199.	137./ 184.	155./ 202.
Solar Engine Cogeneration; Two-Axis Dish, High Eff. Stirling Engines, High- and Low-Temp. Storage, Absorption and Electric Chillers, and Oil Backup.	16	91.4	6.05/ 8.71	5.38/ 7.92	146./ 196.	149./ 198.	137./ 184.	148./ 195.
Solar Engine Cogeneration; Two-Axis Dish, High Eff. Stirling Engines, High- and Low-Temp. Storage, Absorption and Electric Chillers, and Gas Backup.		91.4	6.05/ 8.71	5.38/ 7.92	143./ 192.	146./ 195.	133./ 180.	136./ 183.
Solar Engine Cogeneration; Two-Axis Dish, Low Eff. Stirling Engines, High- and Low-Temp. Storage, Absorption and Electric Chillers, and Oil Backup.		90.4	6.63/ 9.36	5.95/ 8.54	157./ 207.	159./ 210.	147./ 195.	160./ 207.
Solar Engine Cogeneration; Two-Axis Dish, Low Eff. Stirling Engines, High- and Low-Temp. Storage, Absorption and Electric Chillers, and Gas Backup.		90.4	6.63/ 9.36	5.95/ 8.54	152./ 203.	156./ 206.	143./ 191.	147./ 194.

Table 8 (continued)

Summary description of system	Table number	Percent solar	Effective cost of solar energy (¢/kWh)		Levelized monthly cost of energy service			
			No credits	20% ITC	Projection 1 No credits	Projection 2 No credits	Projection 2 20% ITC	Projection 3 20% ITC
100-Percent Solar Engine; Low-Temp. ORCS With River-Water Condenser, 2-Cover Pond, Seasonal Aquifer Storage, and Absorption Chillers	15	100.0	8.85/ 12.33	7.95/ 11.25	207./ 278.	207./ 278.	189./ 256.	189./ 256.
100-Percent Solar Engine; Low-Temp. ORCS With Cooling Tower, 2-Cover Pond, Seasonal Aquifer Storage, and Absorption Chillers		100.0	11.07/ 15.28	9.96/ 13.95	252./ 338.	252./ 338.	230./ 311.	230./ 311.
100-Percent Solar PV Cogeneration; Two-Axis Concentrator With Si Cells (Med. Price), Seasonal Iron-REDOX Electrical and Multitank Low-Temp. Storage, and Absorption Chillers (Minimum Collector Area)		100.0	9.34/ 13.15	8.38/ 12.00	217./ 295.	217./ 295.	198./ 271.	198./ 271
100-Percent Solar PV Cogeneration; Two-Axis Concentrator With Si Cells (Med. Price), Seasonal Iron-REDOX Electrical and Multitank Low-Temp. Storage, and Absorption Chillers (Optimized Collector Area)	17	100.0	7.88/ 11.20	7.05/ 10.19	188./ 255.	188./ 255.	171./ 235.	171./ 235.

Table 9. Summary Description of System: Omaha Community

Summary description of system	Table number	Percent solar	Effective cost of solar energy (¢/kWh)		Levelized monthly cost of energy service			
			No credits	20% ITC	Projection 1 No credits	Projection 2 No credits	Projection 2 20%ITC	Projection 3 20%ITC
OMAHA COMMUNITY								
Conventional systems								
Conv. Heating and Cooling Systems in Each Building; Mixture of Gas/Electric Hot Water, Gas/Heat-Pump/Resistance Heating, and Electric Chilling		0.0	NA/ NA	NA/ NA	98./	133./	133./	235./ NA
Conv. Heating and Cooling Systems in Each Building; All Use Electric Hot Water, Resistance Heating, Electric Cooling, and Utility Electricity		0.0	NA/ NA	NA/ NA	131./	174./	174./ NA	351./ NA
Reference system								
Conv. Heating and Cooling System in Each Building; All Use Electric Hot Water and Electric Cooling; High Rises and Shopping Center Use Resistance Heating, Other Buildings Use Heat Pumps		0.0	NA/ NA	NA/ NA	130./	168./	168./ NA	326./ NA
Systems compared to reference system								
Conv. District Heating; Central Oil Heat and Electric Chilling, and Utility Electricity		34.9	4.49/ 7.49	3.87/ 6.73	127./ 152.	149./ 174.	144./ 168.	237./ 261.
Conv. District Heating; Central Oil Heat and Electric Chilling, and Utility Electricity		34.9	4.49/ 7.49	3.87/ 6.73	116./ 140.	140./ 164.	134./ 158.	203./ 226.
Conv. Engine Cogeneration; Oil-Burning Diesel/ORCS, Absorption and Electric Chillers		55.8	4.98/ 7.70	4.34/ 6.93	134./ 170.	147./ 183.	138./ 173.	197./ 232.
Conv. Engine Cogeneration; Gas-Burning Diesel/ORCS, Absorption and Electric Chillers		55.8	4.98/ 7.70	4.34/ 6.93	114./ 150.	130./ 166.	122./ 156.	137./ 171.
Conv. Engine Cogeneration; Coal Steam Turbines, Absorption and Electric Chillers		45.5	7.72/ 11.85	6.71/ 10.63	139./ 183.	150./ 194.	139./ 181.	173./ 215.
100-Percent Solar Heating; 1-Cover Pond, Seasonal Aquifer Storage, Electric Chillers, and Utility Electricity		60.1	7.29/ 10.63	6.48/ 9.66	174./ 221.	188./ 236.	177./ 222.	237./ 282.

Table 9 (continued)

Summary description of system	Table number	Percent solar	Effective cost of solar energy (¢/kWh)		Levelized monthly cost of energy service			
			No credits	20% ITC	Projection 1 No credits	Projection 2 No credits	Projection 2 20% ITC	Projection 3 20% ITC
100-Percent Solar Heating; Coal Steam Turbines, Absorption and Electric Chillers		60.1	7.01/ 10.03	6.28/ 9.16	170./ 212.	184./ 227.	174./ 215.	234./ 275.
Solar Engine Cogeneration; Heliostats, Steam Turbines, High- and Low-Temp. Storage, Absorption and Electric Chillers, and Coal Backup		67.7	8.83/ 12.89	7.80/ 11.65	188./ 253.	195./ 260.	178./ 240.	198./ 260.
Solar Engine Cogeneration; Heliostats, Steam Turbines With Coal Superheat, High- and Low-Temp. Storage, Absorption and Electric Chillers, and Coal Backup.		65.1	8.41/ 12.37	7.41/ 11.17	177./ 238.	184./ 245.	169./ 227.	191./ 248.
Solar Engine Cogeneration; Two-Axis Dish, High Eff. Stirling Engine, High- and Low-Temp. Storage, Absorption and Electric Chillers, and Oil Backup		87.5	7.42/ 10.68	6.60/ 9.70	197./ 264.	200./ 268.	184./ 248.	200./ 264.
Solar Engine Cogeneration; Two-Axis Dish, High Eff. Stirling Engines, High- and Low-Temp. Storage, Absorption and Electric Chillers, and Gas Backup		87.5	7.42/ 10.68	6.60/ 9.70	191./ 258.	196./ 263.	179./ 243.	183./ 247.
Solar Engine Cogeneration; Two-Axis Dish, Low Eff. Stirling Engines, High- and Low-Temp. Storage, Absorption and Electric Chillers, Oil Backup.		85.8	8.05/ 11.39	7.22/ 10.39	208./ 276.	212./ 280.	195./ 260.	214./ 278.
Solar Engine Cogeneration; Two-Axis Dish, Low Eff. Stirling Engines, High- and Low-Temp. Storage, Absorption and Electric Chillers, Gas Backup.		85.8	8.05/ 11.39	7.22/ 10.39	202./ 269.	207./ 274.	190./ 254.	195./ 259.
100-Percent Heat Engine; Low-Temp. ORCS With River Water Condenser, 2-Cover Pond, Seasonal Aquifer Storage, and Absorption Chillers		100.0	10.45/ 14.29	9.47/ 13.11	280./ 371.	280./ 371.	257./ 343.	257./ 343.
100-Percent Solar Engine; Low-Temp. ORCS With Cooling Tower, 2-Cover Pond, Seasonal Aquifer Storage, Absorption Chillers		100.0	14.30/ 19.54	12.91/ 17.88	371./ 495.	371./ 495.	339./ 456.	339./ 456.

Table 9 (continued)

Summary description of system	Table number	Percent solar	Effective cost of solar energy (¢/kWh)		Levelized monthly cost of energy service			
			No credits	20% ITC	Projection 1 No credits	Projection 2 No credits	Projection 2 20% ITC	Projection 3 20% ITC
100-Percent Solar PV Cogeneration; Two-Axis Concentrator With Si Cells (Med. Price), Seasonal Iron-REDOX Electrical and Multitank Low-Temp. Storage, Absorption Chillers (Minimum Collector Area)		100.0	12.93/ 18.06	11.63/ 16.50	339./ 460.	339./ 460.	308./ 423.	308./ 423.
100-Percent Solar PV Cogeneration; Two-Axis Concentrator With Si Cells (Med. Price), Seasonal Iron-REDOX Electrical and Multitank, Low-Temp. Storage, Absorption Chillers (Optimized Collector Area)		100.0	11.09/ 15.62	9.93/ 14.23	296./ 403.	296./ 403.	268./ 370.	268./ 370.

The first solar systems shown on Tables 8 and 9 are
district heating systems similar to the 100% solar heating
systems examined previously for use with groups of houses.
Table 12 illustrates a system using solar pond collectors
while Table 13 shows a system using flat plate collectors.
In this last system about a quarter of the collectors are
assumed to be mounted on rooftops with the remainder mounted
on racks and foundations at ground level. (It may be
possible, of course, to mount the collectors over parking lots
and roadways avoiding the need for large land areas to be set
aside for use only as collector fields.) The results of analysis of several variants of the systems 12 and 13 are illustrated in Tables 8 and 9. The more expensive of the two pond
systems shown assume 44 $/m^2 for the installed cost of the
collector and 10 ¢/kWh for storage capacity. (This is the
system shown in Table 12.) The less expensive pond system
assumed 30 $/m^2 for the collector and 5 ¢/kWh for storage.
It is assumed that the framing for the pond collectors will
last 30 years but that the plastic lining and covers
(representing approximately half the cost of the installed
collector) will need to be replaced every 10 years. The
more expensive flat plate system assumes a collector cost of
95 $/m^2 (not including installation) and the less expensive
system (the one illustrated in Table 13) assumes a collector
cost of 53 $/m^2. In both flat plate cases storage is assumed
to consist of a series of large buried tanks with an effective
storage cost of 47 ¢/kWh.

The rest of the solar energy systems examined generate
electricity as well as direct heat. The system illustrated
in Table 14 is similar to the coal cogeneration illustrated
in Table 11 except that a heliostat collector field is used to
provide solar heated steam to the turbines when sunlight is
available. Excess solar energy is stored as sensible heat
using a steel ingot storage system. The system also is
provided with a low-temperature storage unit capable of storing
the heat exhausted by the turbine. A possible variant of this
basic system (shown in Tables 8 and 9) would use solar equipment only to boil pressurized water using the coal boiler to
provide superheat.

If low temperature solar collectors and thermal storage
can be constructed at very low cost, it may be preferable to
accept the inefficiencies of a Rankine cycle device operated at
low temperatures. Table 15 illustrates one possible system
based on this concept. The collectors are similar to the low-
cost pond collectors discussed earlier in connection with the
solar district heating system but are assumed to have a somewhat more efficient receiver surface (a selective absorber)
and employ a second plastic cover (only one cover was assumed

Table 10. Albuquerque: Conventional Heat Engine Cogeneration System -- Community Using Oil-Burning Diesel/ORCS, Absorption and Electric Chillers

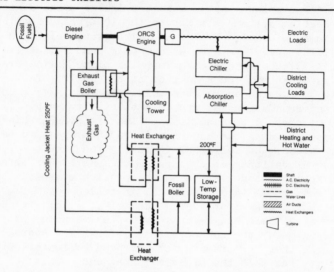

A. ITEMIZED COST OF COMPONENTS

Component	Size	Unit Cost	First Cost (incl O&P)	Annual O&M	Life (yrs)
1. Diesel/ORCS and building	35.2 MW	370 $/kW	$13,000,000	$280,000	30
2. Absorption chillers	7,900 tons	415 $/ton	3,300,000	47,000	30
3. Compression chillers	6,240 tons	260	1,600,000	31,000	30
4. Reserve boiler	108 MW	16 $/kW	1,700,000	69,000	15
5. Heat exchanger	105 MW	4.5 $/kW	470,000	6,800	30
6. 4-pipe thermal distribution	—	—	31,000,000	990,000	30
7. Electric distribution	—	160 $/kW	3,800,000	76,000	30
8. HVAC piping, ducts, and heat exchangers in buildings	—	—	7,900,000	158,000	30
9. Contingencies, parts, indirect costs	—	15%	9,400,000	—	—
10. Interest during construction	—	10%	6,300,000	—	—
TOTAL FOR COMMUNITY			$78,470,000	$1,657,800	
TOTAL PER UNIT			$7,669	$162	

ANNUAL ENERGY FLOWS
(Conventional reference system is cm-2)

	Energy consumed by ref. system	Backup consumed w/ solar/conservation	Energy saved (% of total)
Net Electricity (bought-sold) (MWh/unit)	24.4	0.	100.0
Fuel consumed onsite (MMBtu/unit)	0.	132.	0.
Total energy requirement (bbl crude equiv.)[a]	60.	27.	54.0
Electricity sold to grid annually (MWh,entire building)			0.
Annual peak electricity demand (kW, entire building)			0.

[a] 1 Bbl. crude oil is equivalent to 4.83 mmBtu after 17 percent loss due to refining, transportation, etc. Combined electrical generation, transmission, and distribution efficiency is 29 percent.

[b] The energy cost escalation assumptions are described in detail in chapter II, Volume II, Application of Solar Technology to Today's Energy Needs, Office of Technology Assessment, U. S. Congress, GPO stock number 052-003-00608-1. In all cases, 5.5 percent inflation is assumed.

Table 10 (continued).

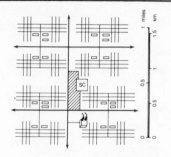

B. LEVELIZED MONTHLY COSTS PER UNIT TO CONSUMER (Dollars)[b,c]

(Conventional reference system is cm-2)

	Escalation of conventional energy costs					
	Constant real energy prices		Energy price escalation I		Energy price escalation II	
1. 1976 Startup						
a. Costs using solar (conservation) system:						
Total with no incentives	122.	(152.)	131.	(161.)	170.	(200.)
Total with 20% ITC	115.	(144.)	123.	(152.)	163.	(191.)
Total with full incentives	105.	(115.)	113.	(124.)	152.	(163.)
b. Costs using conventional reference system	*125.*		*150.*		*253.*	
2. 1985 Startup[d]						
a. Costs using solar (conservation) system:						
(capital related costs)	62.	(92.)	62.	(92.)	62.	(92.)
(operation & maintenance costs)	24.	(24.)	24.	(24.)	24.	(24.)
(fuel bill)	36.	(36.)	50.	(50.)	111.	(111.)
(electric bill)	0.	(0.)	0.	(0.)	0.	(0.)
Total with no incentives	122.	(152.)	135.	(166.)	196.	(227.)
Total with 20% ITC	115.	(144.)	128.	(157.)	189.	(218.)
Total with full incentives	105.	(115.)	118.	(129.)	179.	(190.)
b. Costs using conventional reference system	*125.*		*164.*		*325.*	

C. EFFECTIVE COST OF ENERGY TO CONSUMER

(Conventional reference system is cm-2)

	Type of incentives given					
Levelized cost of solar energy or 'conservation' energy[c]	No incentives		20% ITC		Full incentives	
$/MMBtu primary fuel	4.53	(6.87)	3.98	(6.21)	3.18	(4.01)
¢/kWh electricity	5.33	(8.08)	4.68	(7.31)	3.74	(4.72)

	Escalation of conventional energy costs		
Levelized price paid for conventional energy[b,e]	Constant real energy prices	Energy price escalation I	Energy price escalation II
$/MMBtu primary fuel	*4.09*	*5.14*	*9.45*
¢/kWh electricity	*4.81*	*6.05*	*11.12*

[c] The other costs assume that the energy equipment is owned by the building owners. The equipment in the conventional communities is also owned by the owners of each of the buildings, while in the other communities, it is owned by a municipal utility. In all cases, the parenthesized costs assume ownership by an investor-owned utility using normalized accounting.

[d] "1985 Startup" is the same as "1976 Startup" except that the fuel costs have escalated for 9 years. For ease of comparison with "1976 Startup," 5.5 percent inflation between 1976 and 1985 has been removed.

[e] These levelized prices are computed from the price paid for energy in the reference nonsolar system.

Table 11. Albuquerque: Conventional Heat Engine Cogeneration System -- Community Using Coal Steam Turbines and Absorption and Electric Chillers

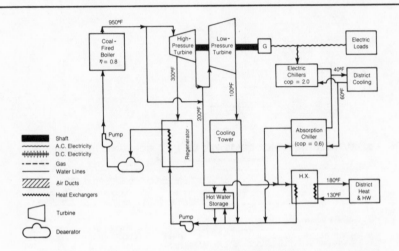

A. ITEMIZED COST OF COMPONENTS

Component	Size	Unit Cost	First Cost (incl O&P)	Annual O&M	Life (yrs)
1. Steam extraction turbine	28.9 MW	360 $/kWe	$10,400,000	$240,000	30
2. Coal boiler (building, fuel stor.)	28.9 MW	490 $/kWe	14,200,000	—	30
3. Absorption chiller	9,200 tons	415 $/ton	3,800,000	55,000	30
4. Compression chiller	5,200 tons	260 $/ton	1,400,000	26,000	30
5. Reserve boiler (coal fired)	95 MW$_t$	100 $/kW$_{th}$	9,500,000	190,000	30
6. Heat exchanger	105 MW	4.5 $/kW	470,000	6,800	30
7. 4-pipe thermal distribution	—	—	31,000,000	990,000	30
8. Electric distribution	—	160 $/kW	3,800,000	76,000	30
9. HVAC piping, ducts, and heat exchangers in buildings	—	—	7,900,000	158,000	30
10. Contingencies, parts, incl. cost	—	15%	12,400,000	0	30
11. Interest during construction	—	10%	8,200,000	0	30
TOTAL FOR COMMUNITY			$103,070,000	$1,741,800	
TOTAL PER UNIT			$10,073	$170	

ANNUAL ENERGY FLOWS
(Conventional reference system is cm-2)

	Energy consumed by ref. system	Backup consumed w/ solar/conservation	Energy saved (% of total)
Net Electricity (bought-sold) (MWh/unit)	24.4	0.	100.0
Fuel consumed onsite (MMBtu/unit)	0.	167.	0.
Total energy requirement (bbl crude equiv.)[a]	60.	35.	41.7
Electricity sold to grid annually (MWh,entire building)			0.
Annual peak electricity demand (kW, entire building)			0.

[a] 1 Bbl. crude oil is equivalent to 4.83 mmBtu after 17 percent loss due to refining, transportation, etc. Combined electrical generation, transmission, and distribution efficiency is 29 percent.

[b] The energy cost escalation assumptions are described in detail in chapter II, Volume II, Application of Solar Technology to Today's Energy Needs, Office of Technology Assessment, U. S. Congress, GPO stock number 052-003-00608-1. In all cases, 5.5 percent inflation is assumed.

Table 11 (continued).

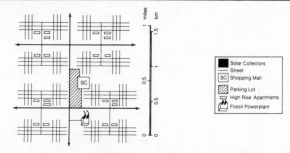

■	Solar Collectors	
	Street	
SC	Shopping Mall	
▨	Parking Lot	
	High Rise Apartments	
	Fossil Powerplant	

B. LEVELIZED MONTHLY COSTS PER UNIT TO CONSUMER (Dollars)[b,c]

(Conventional reference system is cm-2)

	Escalation of conventional energy costs		
	Constant real energy prices	Energy price escalation I	Energy price escalation II
1. 1976 Startup			
a. Costs using solar (conservation) system:			
Total with no incentives....................................	125. (165.)	132. (172.)	152. (192.)
Total with 20% ITC...	116. (153.)	122. (160.)	142. (180.)
Total with full incentives	101. (114.)	108. (120.)	128. (141.)
b. Costs using conventional reference system........	*125.*	*150.*	*253.*
2. 1985 Startup[d]			
a. Costs using solar (conservation) system:			
(capital related costs)..	80. (119.)	80.. (119.)	80. (119.)
(operation & maintenance costs)	25. (25.)	25. (25.)	25. (25.)
(fuel bill)	20. (20.)	31. (31.)	62. (62.)
(electric bill)	0. (0.)	0. (0.)	0. (0.)
Total with no incentives	125. (165.)	136. (175.)	167. (207.)
Total with 20% ITC...	116. (153.)	126. (164.)	157. (195.)
Total with full incentives	101. (114.)	112. (124.)	143. (155.)
b. Costs using conventional reference system........	*125.*	*164.*	*325.*

C. EFFECTIVE COST OF ENERGY TO CONSUMER

(Conventional reference system is cm-2)

	Type of incentives given		
Levelized cost of solar energy or 'conservation' energy[c]	No incentives	20% ITC	Full incentives
$/MMBtu primary fuel......................................	7.80 (11.77)	6.81 (10.58)	5.37 (6.63)
¢/kWh electricity..	9.18 (13.85)	8.02 (12.45)	6.32 (7.81)

	Escalation of conventional energy costs		
Levelized price paid for conventional energy[b,e]	Constant real energy prices	Energy price escalation I	Energy price escalation II
$/MMBtu primary fuel......................................	*4.09*	*5.14*	*9.45*
¢/kWh electricity..	*4.81*	*6.05*	*11.12*

[c] The other costs assume that the energy equipment is owned by the building owners. The equipment in the conventional communities is also owned by the owners of each of the buildings, while in the other communities, it is owned by a municipal utility. In all cases, the parenthesized costs assume ownership by an investor-owned utility using normalized accounting.

[d] "1985 Startup" is the same as "1976 Startup" except that the fuel costs have escalated for 9 years. For ease of comparison with "1976 Startup," 5.5 percent inflation between 1976 and 1985 has been removed.

[e] These levelized prices are computed from the price paid for energy in the reference nonsolar system.

Table 12. Albuquerque: 100-Percent Solar Hot Water & Heating System -- Community Using Single Cover Pond Collector, Seasonal Aquifer Storage, Electric Chillers, and Utility Electricity

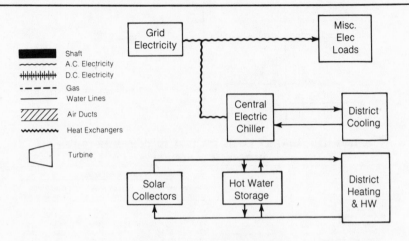

A. ITEMIZED COST OF COMPONENTS

Component	Size	Unit Cost	First Cost (incl O&P)	Annual O&M	Life (yrs)
1. 1-cover shallow pond collector (incl. plumbing and installation).	3.2E5 m²	44 $/m²	*$7,000,000	$190,000	30
			* 7,000,000	0	10
2. Land for collector field (10,000 $/acre, ground cover = ½)	—	5 $/m²	1,600,000	0	30
3. Heat exchangers	1.8E8 kWh	4.5 $/kWh	500,000	5,000	30
4. Electric chiller	12,600 tons	160 $/ton	3,300,000	63,000	30
5. Aquifer thermal storage	1.08E8 kWh	.1 $/kWh	10,800,000	108,000	30
6. 4-pipe thermal distribution system	—	—	31,000,000	990,000	30
7. Electric distribution system	—	160 $/kW	3,800,000	74,000	30
8. HVAC piping, ducts, and heat exchangers in buildings	—	—	7,900,000	158,000	30
9. Contingencies, parts and indirect costs	—	15%	11,000,000	0	30
10. Interest during construction	—	10%	7,400,000	0	30
TOTAL FOR COMMUNITY			$91,300,000	$1,588,000	
TOTAL PER UNIT			$8,923	$155	

* ½ installed collector cost assumed replaced every 10 yrs., with total replacement in 30 yrs.

ANNUAL ENERGY FLOWS
(Conventional reference system is cm-2)

	Energy consumed by ref. system	Backup consumed w/ solar/conservation	Energy saved (% of total)
Net Electricity (bought-sold) (MWh/unit)	24.4	11.0	54.7
Fuel consumed onsite (MMBtu/unit)	0.	0.	0.
Total energy requirement (bbl crude equiv.)[a]	60.	27.	54.7
Electricity sold to grid annually (MWh, entire building)			0.
Annual peak electricity demand (kW, entire building)			38100.0

[a] 1 Bbl. crude oil is equivalent to 4.83 mmBtu after 17 percent loss due to refining, transportation, etc. Combined electrical generation, transmission, and distribution efficiency is 29 percent.

[b] The energy cost escalation assumptions are described in detail in chapter II, Volume II, Application of Solar Technology to Today's Energy Needs, Office of Technology Assessment, U. S. Congress, GPO stock number 052-003-00608-1. In all cases, 5.5 percent inflation is assumed.

Table 12 (continued).

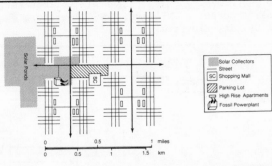

	Solar Collectors
	Street
SC	Shopping Mall
	Parking Lot
	High Rise Apartments
	Fossil Powerplant

B. LEVELIZED MONTHLY COSTS PER UNIT TO CONSUMER (Dollars)[b,c]
(Conventional reference system is cm-2)

	Escalation of conventional energy costs					
	Constant real energy prices		Energy price escalation I		Energy price escalation II	
1. 1976 Startup						
a. Costs using solar (conservation) system:						
Total with no incentives	140.	(175.)	150.	(185.)	190.	(226.)
Total with 20% ITC	131.	(165.)	141.	(175.)	182.	(216.)
Total with full incentives	119.	(131.)	129.	(141.)	169.	(182.)
b. Costs using conventional reference system	125.		150.		253.	
2. 1985 Startup[d]						
a. Costs using solar (conservation) system:						
(capital related costs)	78.	(114.)	78.	(114.)	78.	(114.)
(operation & maintenance costs)	23.	(23.)	23.	(23.)	23.	(23.)
(fuel bill)	0.	(0.)	0.	(0.)	0.	(0.)
(electric bill)	38.	(38.)	54.	(54.)	117.	(117.)
Total with no incentives	140.	(175.)	155.	(191.)	218.	(254.)
Total with 20% ITC	131.	(165.)	147.	(181.)	210.	(244.)
Total with full incentives	119.	(131.)	134.	(147.)	198.	(210.)
b. Costs using conventional reference system	125.		164.		325.	

C. EFFECTIVE COST OF ENERGY TO CONSUMER
(Conventional reference system is cm-2)

	Type of incentives given					
Levelized cost of solar energy or 'conservation' energy[c]	No incentives		20% ITC		Full incentives	
$/MMBtu primary fuel	5.66	(8.38)	5.01	(7.59)	4.06	(4.99)
¢/kWh electricity	6.67	(9.86)	5.90	(8.94)	4.78	(5.87)

	Escalation of conventional energy costs		
Levelized price paid for conventional energy[b,e]	Constant real energy prices	Energy price escalation I	Energy price escalation II
$/MMBtu primary fuel	4.09	5.14	9.45
¢/kWh electricity	4.81	6.05	11.12

[c] The other costs assume that the energy equipment is owned by the building owners. The equipment in the conventional communities is also owned by the owners of each of the buildings, while in the other communities, it is owned by a municipal utility. In all cases, the parenthesized costs assume ownership by an investor-owned utility using normalized accounting.

[d] "1985 Startup" is the same as "1976 Startup" except that the fuel costs have escalated for 9 years. For ease of comparison with "1976 Startup," 5.5 percent inflation between 1976 and 1985 has been removed.

[e] These levelized prices are computed from the price paid for energy in the reference nonsolar system.

Table 13. Albuquerque: 100-Percent Solar Heating and Hot Water System -- Community Using Flat-Plate Collectors (Possible Future Price), Absorption Chillers, Low-Temperature Thermal Storage, and Utility Electricity

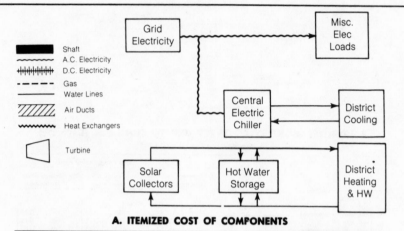

A. ITEMIZED COST OF COMPONENTS

Component	Size	Unit Cost	First Cost (incl O&P)	Annual O&M	Life (yrs)
1. Flat-plate collectors	375,000 m²	53 $/m²	*$9,940,000	0	30
——Collectors @ 50 $/m²			* 9,940,000	0	15
——Transportation @ 3 $/m²					
2. Installation on townhouses and single family houses	106,000 m²	16 $/m²	* 850,000	0	30
			* 850,000	0	15
3. Installation on multifamily residences and shopping ctr.	269,000 m²	41 $/m²	*5,510,000	0	30
——Rack materials @ 15 $/m²			*5,510,000	0	15
——Support columns @ 6 $/m²					
——Installation @ 20 $/m²					
4. Onsite collector plumbing	—	11 $/m² collector	4,100,000	$82,000	30
5. Two-pipe thermal distribution pumps (shared trench)	—	—	14,000,000	270,000	30
6. Low-temperature storage	36E6 kWh	47 $/kWh	16,900,000	34,000	30
7. Onsite equipment (heat exchangers, ductwork, absorption air-conditioning).	—	—	16,200,000	300,000	30
8. Electric distribution system.	—	—	3,800,000	74,000	30
9. Contingencies, parts, & indir.	—	15%	13,200,000	0	—
10. Interest during construction.	—	10%	8,800,000	0	—
TOTAL FOR COMMUNITY			$109,600,000	$760,000	
TOTAL PER UNIT			$10,711	$74	

* ½ installed collector cost assumed replaced in 15 yrs., with total replacement in 30 yrs.

ANNUAL ENERGY FLOWS
(Conventional reference system is cm-2)

	Energy consumed by ref. system	Backup consumed w/ solar/conservation	Energy saved (% of total)
Net Electricity (bought-sold) (MWh/unit)	24.4	8.1	66.9
Fuel consumed onsite (MMBtu/unit)	0.	0.	0.
Total energy requirement (bbl crude equiv.)[b]	60.	20.	66.9
Electricity sold to grid annually (MWh,entire building)			0.
Annual peak electricity demand (kW, entire building)			18850.0

[a] 1 Bbl. crude oil is equivalent to 4.83 mmBtu after 17 percent loss due to refining, transportation, etc. Combined electrical generation, transmission, and distribution efficiency is 29 percent.

[b] The energy cost escalation assumptions are described in detail in chapter II, Volume II, Application of Solar Technology to Today's Energy Needs, Office of Technology Assessment, U. S. Congress, GPO stock number 052-003-00608-1. In all cases, 5.5 percent inflation is assumed.

Table 13 (continued).

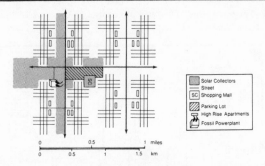

B. LEVELIZED MONTHLY COSTS PER UNIT TO CONSUMER (Dollars)[b,c]
(Conventional reference system is cm-2)

	Escalation of conventional energy costs					
	Constant real energy prices		Energy price escalation I		Energy price escalation II	
1. 1976 Startup						
a. Costs using solar (conservation) system:						
Total with no incentives	128.	(172.)	134.	(178.)	159.	(202.)
Total with 20% ITC	118.	(159.)	124.	(165.)	148.	(189.)
Total with full incentives	102.	(117.)	108.	(123.)	133.	(147.)
b. Costs using conventional reference system	125.		150.		253.	
2. 1985 Startup[d]						
a. Costs using solar (conservation) system:						
(capital related costs)	94.	(137.)	94.	(137.)	94.	(137.)
(operation & maintenance costs)	11.	(11.)	11.	(11.)	11.	(11.)
(fuel bill)	0.	(0.)	0.	(0.)	0.	(0.)
(electric bill)	23.	(23.)	33.	(33.)	71.	(71.)
Total with no incentives	128.	(172.)	138.	(181.)	176.	(219.)
Total with 20% ITC	118.	(159.)	127.	(168.)	165.	(207.)
Total with full incentives	102.	(117.)	112.	(126.)	150.	(164.)
b. Costs using conventional reference system	125.		164.		325.	

C. EFFECTIVE COST OF ENERGY TO CONSUMER
(Conventional reference system is cm-2)

	Type of incentives given					
Levelized cost of solar energy or 'conservation' energy[c]	No incentives		20% ITC		Full incentives	
$/MMBtu primary fuel	4.85	(7.56)	4.19	(6.77)	3.23	(4.13)
¢/kWh electricity	5.71	(8.90)	4.93	(7.96)	3.80	(4.86)

	Escalation of conventional energy costs		
Levelized price paid for conventional energy[b,e]	Constant real energy prices	Energy price escalation I	Energy price escalation II
$/MMBtu primary fuel	4.09	5.14	9.45
¢/kWh electricity	4.81	6.05	11.12

[c] The other costs assume that the energy equipment is owned by the building owners. The equipment in the conventional communities is also owned by the owners of each of the buildings, while in the other communities, it is owned by a municipal utility. In all cases, the parenthesized costs assume ownership by an investor-owned utility using normalized accounting.

[d] "1985 Startup" is the same as "1976 Startup" except that the fuel costs have escalated for 9 years. For ease of comparison with "1976 Startup," 5.5 percent inflation between 1976 and 1985 has been removed.

[e] These levelized prices are computed from the price paid for energy in the reference nonsolar system.

Table 14. Albuquerque: Solar Heat Engine Cogeneration System -- Community Using Steam Turbines, Heliostat Collectors, High- and Low-Temperature Thermal Storage, Absorption and Electric Chillers and Coal Backup

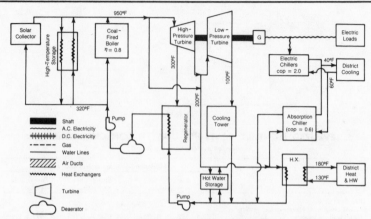

A. ITEMIZED COST OF COMPONENTS

Component	Size	Unit Cost	First Cost (incl O&P)	Annual O&M	Life (yrs)
1. Steam extraction turbine	28.9 MW	360$/kWe	$10,400,000	$230,000	30
2. Coal boiler (building & fuel st.)	28.9 MW	490 $/kWe	14,200,000	0	30
3. Receiving tower & heat exchanger	22.3 MW	150 $/kW	3,300,000	66,000	30
4. Heliostats	167,000 m²	102 $/m²	*8,500,000	170,000	30
——collectors @ 70 $/m²			*8,500,000	170,000	15
——installation @ 30 $/m²					
——transport @ 2 $/m²					
5. Land	—	10 $/m²	1,700,000	0	30
6. Low-temperature storage	256 MWh	0.90 $/MWh	230,000	4,600	30
7. Steel ingot thermal storage	128 MWh	9 $/kWh	1,200,000	23,000	30
8. Absorption chillers	12,480 tons	415 $/ton	5,200,000	75,000	30
9. Compression chillers	5,230 tons	260 $/ton	1,400,000	26,000	30
10. Reserve boiler (coal fired)	93 MW	100 $/kW	9,300,000	190,000	30
11. Heat exchanger	130 MW	4.5 $/kW	500,000	9,100	30
12. HVAC piping, ducts, and heat exchangers in buildings ..	—	—	7,900,000	158,000	30
13. 4-pipe thermal distribution	—	—	31,000,000	990,000	30
14. Electric distribution	—	160 $/kW	3,800,000	76,000	30
15. Contingencies, parts, & indir.	—	15%	16,000,000	0	—
16. Interest during construction.	—	10%	10,700,000	0	—
TOTAL FOR COMMUNITY			$133,830,000	$2,187,700	
TOTAL PER UNIT			$13,080	$214	

* ½ installed collector cost assumed replaced in 15 yrs., with total replacement in 30 yrs.

ANNUAL ENERGY FLOWS
(Conventional reference system is cm-2)

	Energy consumed by ref. system	Backup consumed w/ solar/conservation	Energy saved (% of total)
Net Electricity (bought-sold) (MWh/unit)	24.4	0.	100.0
Fuel consumed onsite (MMBtu/unit)	0.	86.	0.
Total energy requirement (bbl crude equiv.)[a]	60.	18.	70.1
Electricity sold to grid annually (MWh, entire building)			0.
Annual peak electricity demand (kW, entire building)			0.

[a] 1 Bbl. crude oil is equivalent to 4.83 mmBtu after 17 percent loss due to refining, transportation, etc. Combined electrical generation, transmission, and distribution efficiency is 29 percent.

[b] The energy cost escalation assumptions are described in detail in chapter II, Volume II, Application of Solar Technology to Today's Energy Needs, Office of Technology Assessment, U. S. Congress, GPO stock number 052-003-00608-1. In all cases, 5.5 percent inflation is assumed.

Table 14 (continued).

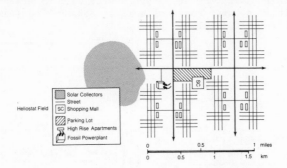

Heliostat Field

Legend	
▓	Solar Collectors
—	Street
SC	Shopping Mall
▨	Parking Lot
	High Rise Apartments
	Fossil Powerplant

```
0        0.5        1  miles
0        0.5    1    1.5  km
```

B. LEVELIZED MONTHLY COSTS PER UNIT TO CONSUMER (Dollars)[b,c]
(Conventional reference system is cm-2)

	Escalation of conventional energy costs					
	Constant real energy prices		Energy price escalation I		Energy price escalation II	
1. 1976 Startup						
a. Costs using solar (conservation) system:						
Total with no incentives	150.	(203.)	154.	(206.)	164.	(216.)
Total with 20% ITC	137.	(187.)	141.	(190.)	151.	(200.)
Total with full incentives	118.	(133.)	121.	(137.)	131.	(147.)
b. Costs using conventional reference system	*125.*		*150.*		*253.*	
2. 1985 Startup[d]						
a. Costs using solar (conservation) system:						
(capital related costs)	108.	(160.)	108.	(160.)	108.	(160.)
(operation & maintenance costs)	32.	(32.)	32.	(32.)	32.	(32.)
(fuel bill)	10.	(10.)	16.	(16.)	32.	(32.)
(electric bill)	0.	(0.)	0.	(0.)	0.	(0.)
Total with no incentives	150.	(203.)	156.	(208.)	172.	(224.)
Total with 20% ITC	137.	(187.)	143.	(192.)	158.	(208.)
Total with full incentives	118.	(133.)	123.	(139.)	139.	(155.)
b. Costs using conventional reference system	*125.*		*164.*		*325.*	

C. EFFECTIVE COST OF ENERGY TO CONSUMER
(Conventional reference system is cm-2)

	Type of incentives given					
Levelized cost of solar energy or 'conservation' energy[c]	No incentives		20% ITC		Full incentives	
$/MMBtu primary fuel	6.72	(9.82)	5.93	(8.87)	4.77	(5.71)
¢/kWh electricity	7.91	(11.56)	6.98	(10.44)	5.62	(6.72)

	Escalation of conventional energy costs		
Levelized price paid for conventional energy[b,e]	Constant real energy prices	Energy price escalation I	Energy price escalation II
$/MMBtu primary fuel	*4.09*	*5.14*	*9.45*
¢/kWh electricity	*4.81*	*6.05*	*11.12*

[c] The other costs assume that the energy equipment is owned by the building owners. The equipment in the conventional communities is also owned by the owners of each of the buildings, while in the other communities, it is owned by a municipal utility. In all cases, the parenthesized costs assume ownership by an investor-owned utility using normalized accounting.

[d] "1985 Startup" is the same as "1976 Startup" except that the fuel costs have escalated for 9 years. For ease of comparison with "1976 Startup," 5.5 percent inflation between 1976 and 1985 has been removed.

[e] These levelized prices are computed from the price paid for energy in the reference nonsolar system.

Table 15. Albuquerque: 100-Percent Solar Heat Engine System -- Community Using Low-Temperature ORCS with River-Water Condenser, Two-Cover Pond Collectors, Seasonal Aquifer Storage, and Absorption Chillers

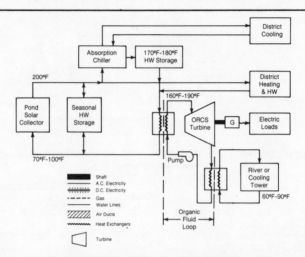

A. ITEMIZED COST OF COMPONENTS

Component	Size	Unit Cost	First Cost (incl O&P)	Annual O&M	Life (yrs)
1. 2-cover shallow pond collector—sel surf. (incl. plumbing & installation).	1.5 E6m²	40 $/m²	*$30,000,000	0	30
			* 30,000,000	$900,000	10
2. Land for collector field (10,000 $/acre, ground cover - ½).	—	5$/m²	7,500,000	0	30
3. Freon engine	1.885E4 kW	707 $/kW	13,300,000	133,000	30
4. Heat exchangers	2.1E5 kW	4.5 $/kW	900,000	9,000	30
5. Absorption chiller	12,600 tons	415 $/ton	5,200,000	76,000	30
6. 170° F water storage tank	4E5 kWh	0.7 $/kWh	300,000	3,000	30
7. Aquifer thermal storage	2.6E8 kWh	0.05 $/kWh	13,060,000	130,000	30
8. 4-pipe thermal distribution system	—	—	31,000,000	990,000	30
9. Electric distribution system	—	160 $/kW	3,800,000	74,000	30
10. HVAC piping, ducts, and heat exchangers in buildings ..	—	—	7,900,000	158,000	30
11. Contingencies, parts, & indirect costs	—	15%	21,400,000	0	30
12. Interest during construction	—	10%	14,300,000	0	30
TOTAL FOR COMMUNITY			$178,600,000	$2,473,000	
TOTAL PER UNIT			$17,455	$242	

* ½ Installed collector cost assumed replaced every 10 yrs., with total replacement in 30 yrs.

ANNUAL ENERGY FLOWS
(Conventional reference system is cm-2)

	Energy consumed by ref. system	Backup consumed w/ solar/conservation	Energy saved (% of total)
Net Electricity (bought-sold) (MWh/unit)	24.4	0.	100.0
Fuel consumed onsite (MMBtu/unit)	0.	0.	0.
Total energy requirement (bbl crude equiv.)[a]	60.	0.	100.0
Electricity sold to grid annually (MWh,entire building)			0.
Annual peak electricity demand (kW, entire building)			0.

[a] 1 Bbl. crude oil is equivalent to 4.83 mmBtu after 17 percent loss due to refining, transportation, etc. Combined electrical generation, transmission, and distribution efficiency is 29 percent.

[b] The energy cost escalation assumptions are described in detail in chapter II, Volume II, Application of Solar Technology to Today's Energy Needs, Office of Technology Assessment, U. S. Congress, GPO stock number 052-003-00608-1. In all cases, 5.5 percent inflation is assumed.

Table 15 (continued).

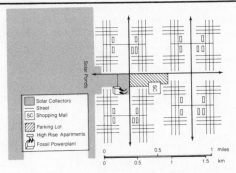

Legend:
- Solar Collectors
- Street
- SC Shopping Mall
- Parking Lot
- High Rise Apartments
- Fossil Powerplant

Solar Ponds

B. LEVELIZED MONTHLY COSTS PER UNIT TO CONSUMER (Dollars)[b,c]
(Conventional reference system is cm-2)

	Escalation of conventional energy costs		
	Constant real energy prices	Energy price escalation I	Energy price escalation II
1. 1976 Startup			
a. Costs using solar (conservation) system:			
Total with no incentives	207. (278.)	207. (278.)	207. (278.)
Total with 20% ITC	189. (256.)	189. (256.)	189. (256.)
Total with full incentives	162. (183.)	162. (183.)	162. (183.)
b. Costs using conventional reference system	125.	150.	253.
2. 1985 Startup[d]			
a. Costs using solar (conservation) system:			
(capital related costs)	171. (242.)	171. (242.)	171. (242.)
(operation & maintenance costs)	36. (36.)	36. (36.)	36. (36.)
(fuel bill)	0. (0.)	0. (0.)	0. (0.)
(electric bill)	0. (0.)	0. (0.)	0. (0.)
Total with no incentives	207. (278.)	207. (278.)	207. (278.)
Total with 20% ITC	189. (256.)	189. (256.)	189. (256.)
Total with full incentives	162. (183.)	162. (183.)	162. (183.)
b. Costs using conventional reference system	125.	164.	325.

C. EFFECTIVE COST OF ENERGY TO CONSUMER
(Conventional reference system is cm-2)

	Type of incentives given		
Levelized cost of solar energy or 'conservation' energy[c]	No incentives	20% ITC	Full incentives
$/MMBtu primary fuel	7.52 (10.47)	6.76 (9.56)	5.64 (6.50)
¢/kWh electricity	8.85 (12.33)	7.95 (11.25)	6.64 (7.65)

	Escalation of conventional energy costs		
Levelized price paid for conventional energy[b,e]	Constant real energy prices	Energy price escalation I	Energy price escalation II
$/MMBtu primary fuel	4.09	5.14	9.45
¢/kWh electricity	4.81	6.05	11.12

[c] The other costs assume that the energy equipment is owned by the building owners. The equipment in the conventional communities is also owned by the owners of each of the buildings, while in the other communities, it is owned by a municipal utility. In all cases, the parenthesized costs assume ownership by an investor-owned utility using normalized accounting.

[d] "1985 Startup" is the same as "1976 Startup" except that the fuel costs have escalated for 9 years. For ease of comparison with "1976 Startup," 5.5 percent inflation between 1976 and 1985 has been removed.

[e] These levelized prices are computed from the price paid for energy in the reference nonsolar system.

Table 16. Albuquerque: Solar Heat Engine Cogeneration System -- Community Using Two-Axis Parabolic Concentrators, High Efficiency Stirling Engines, High- and Low-Temperatures, High Storage, Absorption and Electric Chillers, and Oil Backup

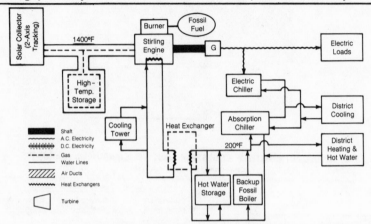

A. ITEMIZED COST OF COMPONENTS

Component	Size	Unit Cost	First Cost (incl O&P)	Annual O&M	Life (yrs)
1. 2-axis tracking collector	167,000 m²	164 $/m²	*$13,700,000	$167,000	30
——tracking unit @ 115 $/m²			* 13,700,000	167,000	15
——installation @ 46 $/m²					
——shipping @ 3 $/m²					
2. High temperature (1400 °F) thermal storage (MgFe)	3 × 10⁵ kWh	34 $/kWh	10,200,000	200,000	30
3. Land (10,000 $/acre ground cover = 1/4)	—	10 $/m²	1,700,000	0	30
4. Low-temperature (hot water) thermal storage	3×10⁶ kWh	0.40 $/kWh	1,200,000	24,000	30
5. Stirling engine (47% efficient)	30.7 MW	135 $/kW	4,100,000	170,000	15
——engine, generator, switch gear controls, and installation @ 100 $/kW					
——heat pipe @ 5 $/kW					
——collector attachment devices @ 30 $/kw					
6. Collector pipefield (incl. pumps and controls)	—	18 $/m²	3,000,000	60,000	30
7. Heat exchangers	65.3 MW	4.5 $/kW$_t$	290,000	5,800	30
8. Absorption chiller	12,600 tons	415 $/ton	5,200,000	76,000	30
9. Compression chiller	9,310 tons	260 $/ton	2,400,000	47,000	30
10. Reserve boiler	109 MW	16 $/kW	1,700,000	68,000	15
11. 4-pipe thermal distribution system	—	160 $/kW	31,000,000	990,000	30
12. Electric distribution system	—	—	3,800,000	74,000	30
13. HVAC piping, ducts, and heat exchangers in buildings	—	—	7,900,000	158,000	30
14. Contingencies, parts, and indirect costs	—	15%	15,100,000	0	30
15. Interest during construction	—	10%	10,100,000	0	30
TOTAL FOR COMMUNITY			$125,090,000	$2,206,800	
TOTAL PER UNIT			$12,225	$216	

* ½ installed collector cost assumed replaced in 15 yrs., with total replacement in 30 yrs.

ANNUAL ENERGY FLOWS (Conventional reference system is cm-2)

	Energy consumed by ref. system	Backup consumed w/ solar/conservation	Energy saved (% of total)
Net Electricity (bought-sold) (MWh/unit)	24.4	0.	100.0
Fuel consumed onsite (MMBtu/unit)	0.	25.	0.
Total energy requirement (bbl crude equiv.)[a]	60.	5.	91.4
Electricity sold to grid annually (MWh,entire building)			0.
Annual peak electricity demand (kW, entire building)			0.

[a] 1 Bbl. crude oil is equivalent to 4.83 mmBtu after 17 percent loss due to refining, transportation, etc. Combined electrical generation, transmission, and distribution efficiency is 29 percent.

[b] The energy cost escalation assumptions are described in detail in chapter II, Volume II, Application of Solar Technology to Today's Energy Needs, Office of Technology Assessment, U. S. Congress, GPO stock number 052-003-00608-1. In all cases, 5.5 percent inflation is assumed.

Table 16 (continued).

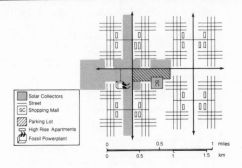

Solar Collectors
Street
SC Shopping Mall
Parking Lot
High Rise Apartments
Fossil Powerplant

0   0.5   1   miles
0   0.5   1   1.5   km

B. LEVELIZED MONTHLY COSTS PER UNIT TO CONSUMER (Dollars)[b,c]

(Conventional reference system is cm-2)

	Escalation of conventional energy costs					
	Constant real energy prices		Energy price escalation I		Energy price escalation II	
1. 1976 Startup						
a. Costs using solar (conservation) system:						
Total with no incentives	146.	(196.)	148.	(198.)	155.	(205.)
Total with 20% ITC	134.	(181.)	136.	(183.)	143.	(190.)
Total with full incentives	116.	(132.)	118.	(134.)	125.	(141.)
b. Costs using conventional reference system	125.		150.		253.	
2. 1985 Startup[d]						
a. Costs using solar (conservation) system:						
(capital related costs)	108.	(157.)	108.	(157.)	108.	(157.)
(operation & maintenance costs)	32.	(32.)	32.	(32.)	32.	(32.)
(fuel bill)	7.	(7.)	9.	(9.)	21.	(21.)
(electric bill)	0.	(0.)	0.	(0.)	0.	(0.)
Total with no incentives	146.	(196.)	149.	(198.)	160.	(210.)
Total with 20% ITC	134.	(181.)	137.	(184.)	148.	(195.)
Total with full incentives	116.	(132.)	119.	(135.)	130.	(146.)
b. Costs using conventional reference system	125.		164.		325.	

C. EFFECTIVE COST OF ENERGY TO CONSUMER

(Conventional reference system is cm-2)

	Type of incentives given					
Levelized cost of solar energy or 'conservation' energy[c]	No incentives		20% ITC		Full incentives	
$/MMBtu primary fuel	5.14	(7.40)	4.57	(6.73)	3.75	(4.48)
¢/kWh electricity	6.05	(8.71)	5.38	(7.92)	4.42	(5.27)

	Escalation of conventional energy costs		
Levelized price paid for conventional energy[b,e]	Constant real energy prices	Energy price escalation I	Energy price escalation II
$/MMBtu primary fuel	4.09	5.14	9.45
¢/kWh electricity	4.81	6.05	11.12

[c] The other costs assume that the energy equipment is owned by the building owners. The equipment in the conventional communities is also owned by the owners of each of the buildings, while in the other communities, it is owned by a municipal utility. In all cases, the parenthesized costs assume ownership by an investor-owned utility using normalized accounting.

[d] "1985 Startup" is the same as "1976 Startup" except that the fuel costs have escalated for 9 years. For ease of comparison with "1976 Startup," 5.5 percent inflation between 1976 and 1985 has been removed.

[e] These levelized prices are computed from the price paid for energy in the reference nonsolar system.

Table 17. Albuquerque: 100-Percent Solar Photovoltaic Cogeneration System -- Community Using Two-Axis Concentrator with Silicon Cells (Medium Price), Seasonal Iron-REDOX Electrical and Multitank Low-Temperature Thermal Storage, and Absorption Chillers (Optimized Collector Area)

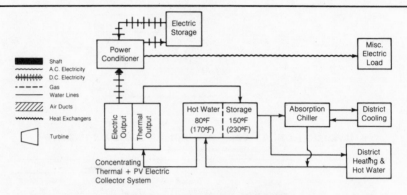

A. ITEMIZED COST OF COMPONENTS

Component	Size	Unit Cost	First Cost (incl O&P)	Annual O&M	Life (yrs)
1. 2-axis tracking collector (with Si solar cells)	4.5E5 m²	114 $/m²	*$25,700,000	$450,000	30
——solar cells @ 5 $/m²			* 25,700,000	450,000	15
——tracking collector 80 $/m²					
——transportation @ 3 $/m²					
——installation @ 26 $/m²					
2. Land for collector field (10,000 $/acre ground cover=¼).	—	10 $/m²	4,500,000	0	30
3. Low-temperature (hot water) storage—multitank	2.5E6 kWh	.60 $/kWh	1,500,000	15,000	30
4. Power conditioner	2E4 kW	85$/kW	1,700,000	17,000	30
5. Collector pipefield (incl. pumps & controls)	—	18$/m²	8,100,000	81,000	30
6. Heat exchangers	2.5E5 kWt	4.5 $/kWt	1,100,000	11,000	30
7. Absorption chiller	12,480 tons	415 $/ton	5,200,000	75,000	30
8. Battery—Iron Redox	1.5E6 kWh	9.4 $/kWh	14,100,000	141,000	15
9. Reserve boiler	1.14E5	16 $/kW	1,800,000	9,000	30
10. Battery building & installation	—	1.1 $/kWh	1,700,000	17,000	30
11. 4-pipe thermal distribution system	—	—	31,000,000	990,000	30
12. Electric distribution system	—	160 $/kW	3,800,000	74,000	30
13. HVAC piping, ducts, and heat exchangers in buildings ..	—	—	7,900,000	158,000	30
14. Contingencies, parts, & indirect costs	—	15%	20,000,000	0	30
15. Interest during construction	—	10%	13,400,000	0	30
TOTAL FOR COMMUNITY			$167,200,000	$2,488,000	
TOTAL PER UNIT			$16,341	$243	

*½ installed collector cost assumed replaced in 15 yrs., with total replacement in 30 yrs.

ANNUAL ENERGY FLOWS
(Conventional reference system is cm-2)

	Energy consumed by ref. system	Backup consumed w/ solar/conservation	Energy saved (% of total)
Net Electricity (bought-sold) (MWh/unit)	24.4	0.	100.0
Fuel consumed onsite (MMBtu/unit)	0.	0.	0.
Total energy requirement (bbl crude equiv.)[a]	60.	0.	100.0
Electricity sold to grid annually (MWh,entire building)			0.
Annual peak electricity demand (kW, entire building)			0.

[a] 1 Bbl. crude oil is equivalent to 4.83 mmBtu after 17 percent loss due to refining, transportation, etc. Combined electrical generation, transmission, and distribution efficiency is 29 percent.

[b] The energy cost escalation assumptions are described in detail in chapter II, Volume II, Application of Solar Technology to Today's Energy Needs, Office of Technology Assessment, U. S. Congress, GPO stock number 052-003-00608-1. In all cases, 5.5 percent inflation is assumed.

Table 17 (continued).

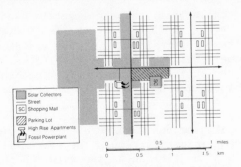

Solar Collectors
Street
SC Shopping Mall
Parking Lot
High Rise Apartments
Fossil Powerplant

B. LEVELIZED MONTHLY COSTS PER UNIT TO CONSUMER (Dollars)[b,c]
(Conventional reference system is cm-2)

	Escalation of conventional energy costs		
	Constant real energy prices	Energy price escalation I	Energy price escalation II
1. 1976 Startup			
a. Costs using solar (conservation) system:			
Total with no incentives	188. (255.)	188. (255.)	188. (255.)
Total with 20% ITC	171. (235.)	171. (235.)	171. (235.)
Total with full incentives	146. (167.)	146. (167.)	146. (167.)
b. Costs using conventional reference system	125.	150.	253.
2. 1985 Startup[d]			
a. Costs using solar (conservation) system:			
(capital related costs)	152. (219.)	152. (219.)	152. (219.)
(operation & maintenance costs)	36. (36.)	36. (36.)	36. (36.)
(fuel bill)	0. (0.)	0. (0.)	0. (0.)
(electric bill)	0. (0.)	0. (0.)	0. (0.)
Total with no incentives	188. (255.)	188. (255.)	188. (255.)
Total with 20% ITC	171. (235.)	171. (235.)	171. (235.)
Total with full incentives	146. (167.)	146. (167.)	146. (167.)
b. Costs using conventional reference system	125.	164.	325.

C. EFFECTIVE COST OF ENERGY TO CONSUMER
(Conventional reference system is cm-2)

	Type of incentives given		
Levelized cost of solar energy or 'conservation' energy[c]	No incentives	20% ITC	Full incentives
$/MMBtu primary fuel	6.70 (9.52)	5.99 (8.66)	4.95 (5.82)
¢/kWh electricity	7.88 (11.20)	7.05 (10.19)	5.83 (6.85)

	Escalation of conventional energy costs		
Levelized price paid for conventional energy[b,e]	Constant real energy prices	Energy price escalation I	Energy price escalation II
$/MMBtu primary fuel	4.09	5.14	9.45
¢/kWh electricity	4.81	6.05	11.12

[c] The other costs assume that the energy equipment is owned by the building owners. The equipment in the conventional communities is also owned by the owners of each of the buildings, while in the other communities, it is owned by a municipal utility. In all cases, the parenthesized costs assume ownership by an investor-owned utility using normalized accounting.

[d] "1985 Startup" is the same as "1976 Startup" except that the fuel costs have escalated for 9 years. For ease of comparison with "1976 Startup," 5.5 percent inflation between 1976 and 1985 has been removed.

[e] These levelized prices are computed from the price paid for energy in the reference nonsolar system.

in the heating collectors.) The collectors used in the
heating system are assumed to cost 40 $/m^2. Several variants of
the basic low-temperature Rankine system are shown in Tables 8
and 9. The lower cost system is assumed to use 15° C river
water to cool the engine condenser while the other system uses
a cooling tower assumed to be capable of providing 32° C
cooling for the engine condenser.

Table 16 illustrates a solar electric device operating at
the opposite temperature extreme. The system illustrated
assumes the use of a Stirling engine with 47% thermal
efficiency mounted at the focus of a fully tracking parabolic
dish capable of supplying heat to the hot side of the engine
at 760° C. The installed cost of these collectors is assumed
to be 205 $/m^2. A system using engines assumed to be 32%
efficient was also examined and the results summarized in
Tables 8 and 9.

The final system examined is similar to the Stirling
engine system just described except that electrical
conversion is accomplished using silicon photovoltaic cells
designed to be operated in the high light concentrations
produced in a two-axis tracking collector (see Table 17).
The installed collector is assumed to cost 143 $/m^2 (including
the photovoltaic cells). The cooling water used to keep the
cells from overheating is used to provide district heating.
It should also be noticed that the results of the Stirling
engine analysis discussed in the previous paragraph can also
be used to estimate the cost of concentrating photovoltaic
systems using very efficient cells. The additional cost of
mounting a heavy engine instead of a photovoltaic device
at the focus of the collector would approximately cancel the
difference in the cost of the generating units.

VI. Conclusions

No easy interpretation of the results of the analysis
is possible. With a tax credit or with municipal utility
financing, a number of very large solar systems are able to
compete with conventional utility costs in Albuquerque and are
surprisingly close to the cost of the fossil cogeneration
systems. As expected, the solar systems are somewhat less
attractive in Omaha where the solar energy resource is smaller.

Since the thermal distribution system adds considerably
to the costs, it is possible that the community chosen for
analysis is too large for an optimum solar community system.
Much more analysis would be required, however, to determine
the optimum size versus density for a solar community.

Some of the systems examined in this study are admittedly speculative and many of them would be difficult to retrofit into an existing community. Given the turnover of existing energy equipment, and the rapid introduction of novel energy devices which can be expected during the next few years, however, a transition to many of the systems outlined here may not be much more ambitious than the changes which would be expected without any attempt to develop a coordinated community approach.

At a minimum, a better understanding of the operation of integrated energy systems is needed to determine the characteristics desired of the system components which are being developed and to provide better guidance for public policy in energy. It would be difficult to construct a package of incentives designed to encourage solar energy usage in a socially preferred manner without adequately understanding which of many possible solutions would be preferred on technical grounds if no other constraints existed.

Appendix A

Economic Methodology

Each system treated was costed on an individual component or subsystem basis with the first cost, annual O & M and the lifetime specified as illustrated in part A of Tables 10-17.

The costs of operating this equipment can be divided into four broad categories:

1) <u>Capital Costs.</u> These include the cost of paying investors for their funds, and any taxes and insurance which must be paid on tangible property, less any tax benefits provided such as the deduction for interest paid. In most cases, all of these costs will be directly proportional to the initial cost of the system.

2) <u>Operating and Maintenance (O&M) Costs.</u> These include costs of keeping equipment in repair, paying operators, etc. They do not include fuel costs.

3) <u>Energy Costs.</u> These include the price paid for all fossil fuels and electricity used by the equipment. In cases where energy can be sold to a utility, the owner's energy costs are reduced by the amount of income received from this source.

4) <u>Replacement Costs.</u> These include purchase of those large pieces of equipment which wear out before the rest of the system and must be replaced.

Most of the differences between owners are reflected, in the costs of capital, since this represents differences in the expected rate of return and in tax status. However, the component of the average cost of energy to the final consumer, which is traceable to capital costs, is proportional to the initial cost regardless of ownership. This capital cost is written in the following form:

$$\begin{bmatrix} \text{average capital charges perceived} \\ \text{by the final consumer of energy} \end{bmatrix} = k_1 \times \begin{bmatrix} \text{initial cost of} \\ \text{system} \end{bmatrix}$$

The constant in this equation, k_1, is called the "Capital Multiplier." It represents the ratio between the portion of annual consumer costs attributable to capital-related expenses and the initial capital cost. The figures implicitly assume inflation, since the interest rates and rates of return expected reflect actual market rates.

The routine Operating and Maintenance (O&M) costs of a system are written in the following form:

$$\begin{bmatrix} \text{average O\&M costs perceived} \\ \text{by energy consumer} \end{bmatrix} = k_2 \times \begin{bmatrix} \text{O\&M cost in first year} \\ \text{of the system's opera-} \\ \text{tion} \end{bmatrix}$$

where the constant k_2 depends on the consumer's discount rate, the life expectancy of the system, and on the average rate of inflation. It is assumed that operating costs do not change in constant dollars for the life of the system. This represents a considerable simplification of real cases, since the costs of maintaining and repairing real systems increase as the system ages. The approximation used here is necessary, however, since it is difficult or impossible to estimate the maintenance schedules reliably, particularly for untested or hypothetical systems.

The fuel costs are written in the following form:

$$\begin{bmatrix} \text{average energy costs} \\ \text{perceived by consumer} \end{bmatrix} = k_3 \times \begin{bmatrix} \text{energy costs in the first} \\ \text{year of the system's} \\ \text{operation} \end{bmatrix}$$

where k_3 depends on the life expectancy of the system, the consumer's discount rates, and assumptions about the rate at which energy from conventional sources will increase in price.

Table A-1. Values of the Constants k_1, k_2, and k_3 Used in the Analysis

Owner	k_1	k_1 With 20% Solar ITC	k_2
Homeowner, New	0.094	0.075	1.777
Private Utility	0.142	0.126	1.777
Municipal Utility	0.095	0.081	1.777

	Price Projection 1 (No Escalation)	Price Projection 2 (Escalation I)	Price Projection (Escalation II)
(1985 Startup)			
k_3 (oil)	1.098	1.497	3.341
k_3 (gas)	1.098	2.264	3.341
k_3 (coal)	1.098	1.658	3.341
k_3 (electricity)	1.098	1.540	3.341

The replacement costs are somewhat more complex, since they depend on the number and schedule of replacements.

Using the terms defined here, the average annual cost of energy to the ultimate customer of that energy (which is called PRICE) can be written in the following form:

PRICE=k_1x(initial price of equipment)+k_2x(initial annual O&M costs)+k_3x(initial annual energy costs) +(replacement costs).

The levelized monthly cost to the consumer is then one-twelfth of PRICE. Present value analysis was used to determine each of these terms and the values used in this paper are shown in Table A-1.

Appendix B

Guide to Tables 8-9

The summary tables provide the following information about each system examined:

· A brief descriptive title.

· A table number that indicates which of Tables 10-17 describes that system in more detail (no detailed table is provided for systems where this column is left blank).

· The percentage of the energy used by the reference system that is supplied by the solar energy system. If an energy conservation system is employed that does not use solar energy, this number represents the percentage energy saving. The formula for this percentage (P_s) is as follows:

$$P_s = 100 \times \frac{E_r - E_t}{E_r}$$

Subscript "r" refers to the reference system, subscript "t" refers to the test system, i.e., the system being compared with the reference system.

· The effective cost of solar energy with and without an investment tax credit. (When a conservation system is shown which does not use solar energy, this cost reflects the effective cost of saving energy using the conservation system.)

The formula for effective energy cost (EC) is as follows:

$$EC = \frac{C_r + OM_r + CR_r - C_t - OM_t - CR_t}{E_r - E_t}$$

C = levelized annual capital costs (including financing charges, taxes, and insurance)

OM = levelized annual operating and maintenance costs (excluding purchases of electricity and fuels)

CR = levelized cost of replacements.

• The levelized monthly energy costs make four different assumptions about the kinds of tax credits given and the cost of conventional energy.

The tables indicate the costs resulting from financing by a municipal utility. This cost is paired with a cost that would result if the additional solar or conservation equipment were owned instead by a privately owned utility. Here, it is assumed that the owner other than the private utility owns a share equal in value to the cost of the reference energy system (i.e., the backup system, in most cases).

Appendix C

Guide to Tables 10-17

Each table describes a single energy system designed to serve the community. The pages are divided into three parts:

• A diagram of the layout of the community.

• An energy flow-diagram indicating the way in which collectors, storage devices, engines, and energy-consuming devices are combined to meet the energy demands of the buildings.

• A set of three tables providing details about the costs and performance assumed for the system.

-- Table A provides an itemized cost list of all components used in the system. It includes an estimate of the first cost, the annual operating and maintenance costs (exclusive of purchased energy) which are charged during the first year of the system's operation, and the expected lifetime of the component (rounded to 10, 15, or 30 years).

The second part of the table indicates the amount of nonsolar energy (electricity or fossil fuel) that was purchased to provide backup for the solar energy system.

-- Table B provides estimates of the levelized monthly costs for a system that begins operation in 1976, and also for a similar system that would begin operations in 1985. The only difference between the 1976 cases and the 1985 cases is that the conventional energy prices in the 1985 cases have escalated to a higher level by the startup date. For ease of comparison with the 1976 cases, prices have not been inflated between 1976 and 1985 for the 1985 cases (i.e., all cases start in 1976 dollars, and costs inflate at 5.5 percent in each succeeding year). The 1976 costs are shown only for reference purposes; they are not meant to suggest that all technologies examined were available in 1976. The levelized cost of the conventional community with heat pumps is also shown for comparison.

The costs achieved with a 20-percent investment tax credit (ITC) would also be reached with low-interest loans and other incentives. The table below shows the interest rate of a loan which would result in the same annual capital costs (k_1) as a 20-percent ITC for several different equipment owners.

The costs achieved with "full incentives" assumed a combination of 20-percent investment tax credit, 3-year straight-line depreciation, and exemption from property taxes. These cost reductions could also be reached with tax credits or other incentives.

Table A-2. Interest Rates Corresponding to Conventional Rates 20% ITC

Owner	k_1 20%ITC	Loan interest (95% financed)
Homeowner, new construction ..	0.075	0.065
Homeowner, retrofit	0.092	0.043
Real estate owner	0.067	0.070
Industry (20% internal rate of return)	0.239	0.274
Municipal utility	0.081	0.041
Private utility	0.126	0.045

-- Table C provides an estimate of the effective cost of solar (or conservation) energy computed using the technique described in Appendix B. The cost of conventional electricity and fuels levelized over the same time interval are provided for comparison.

References

(1) The "baseline" residential energy price projection from BNL was supplied by Eric Hirst, Oak Ridge National Laboratory, July 1976. There is no single "standard" set of BNL energy price projections. A number of different scenarios have been run, yielding different results, and BNL is constantly updating its projections as new data becomes available.

(2) Application of Solar Technology to Today's Energy Needs, Volumes I and II. Office of Technology Assessment, U. S. Congress, GPO stock numbers 052-003-00539-5 and 052-003-00608-1.

(3) A more detailed description of the economic methodology and itemized costs for each system mentioned in this paper is given in Volume II of (2).

COMMENTS

Larry S. James (Washington D.C.): Why were Albuquerque, New Mexico, results presented as opposed to other possible locations ?

Author's Reply:

Albuquerque was presented as an example of a relatively favorable location for using solar energy. The insolation received is high, but gas and electricity are priced below the national average. Omaha was presented as an example of a relatively unfavorable location, which combines average insolation with gas and electric prices which are well below average.

8

The Effect of Solar Heating and Hot Water Design Strategies on Future Market Acceptance

Gerald E. Bennington and Peter C. Spewak

ABSTRACT

This paper presents current cost/performance relations for solar heating and hot water systems and projects probable cost/performance relationships which may be achieved with currently envisioned technology. Based on these projections, three alternative design objectives are postulated: 1) minimum $/f^2 per given efficiency; 2) maximize efficiency within cost constraints; 3) maximize efficiency within size constraints. The results show that alternative two yields the lowest cost on the basis of initial system cost; alternative three provides the lowest cost in terms of dollars per million Btu. The cost/performance results were then used with the MITRE/Metrek market penetration algorithm. This algorithm estimates consumer's propensity to buy solar systems based on the maturity of the technology, the initial cost of the solar system, and the annual savings which may be realized through the use of solar systems. It was determined that alternative two would have the largest market penetration impact even though system three provided lower cost (and higher annual savings).

INTRODUCTION

Solar heating and cooling of buildings (SHACOB) is still in a precommercialization phase at the current time in the United States. Although the technology is well understood, there is much room for improvement in terms of cost and performance. Because the solar industry itself is at best still in its adolescence, the opportunity still exists to shape the potential technological development in such a way as to maximize the appeal for solar heating, cooling, and hot water and thus maximize its use. In developing new solar

systems which will be the most appealing to potential consumers, the designer has three options:

● Minimize cost
● Maximize efficiency
● Trade off one of the above for the other in a cost/
  efficiency compromise.

In order to make a choice, one must consider where the
technologies stand today, what their potential improvements
are, and how these improvements may impact the consumer
appeal of solar systems.

This paper discusses current cost/performance characteristics and presents projections of what may be realistically
expected in terms of ultimate cost/performance relationships
given the advent of an ambitious industry/government research
and development program. Finally, given these potential
cost/performance relationships and some knowledge as to how
consumers view solar systems in terms of initial cost and
performance related savings, three alternative cost/
performance design development strategies are hypothesized
and evaluated in terms of potential market penetration.

SHACOB MODULES SUBJECT TO COST REDUCTION

Cost reductions are expected in the following modules of
solar heating, cooling and hot water systems:

Collector Module

This is the solar energy collection and conversion device
in the SHACOB system. The predominant types of collector in
the market place at this time are the flat plate and the
evacuated-tube nonconcentrating collectors.[1] In this module,
solar radiation strikes a blackened or selectively surfaced
absorber and is converted to thermal energy. A glazing system
allows sunlight to enter the collector and reduce undesired
cooling of the absorber due to wind, reradiation and convection. The insulated collector housing retards the escape
of thermal energy before it is removed from the absorber by
the gaseous or liquid heat transfer medium circulating in
contact with the absorber.

Estimates for cost reduction are based upon The MITRE
Corporation, Metrek Division's analysis of materials and
labor requirements for mass production facilities. It is
expected, based upon this analysis, that a cost decrease of
$4 to $5 per square foot of flat plate collector may be
witnessed due to automation of production facilities,

development of new manufacturing techniques and substitution materials.

Evacuated tube collector costs may decrease by as much as $14 per square foot due to the effects of automation. These estimates are summarized in Table I.

Controls and Storage

The control module is the "brain" of the SHACOB system which regulates the flow of heat transfer medium (hence thermal energy) through the system. Energy may be directed to storage or to immediate use. Control is also exercised over auxiliary heat sources depending upon the solar system's ability to meet the demand for heat by the system's load.

Storage of thermal energy is accomplished primarily as sensible heat increases in solids and liquids stored in insulated bins or tanks.[2]

The cost of control systems is expected to decrease approximately 40 percent and the cost of storage systems 20 to 40 percent. Control system cost decreases are based on the assumption that electrical/electronic circuitry costs will decrease by 60 percent through mass production and that the cost of electromechanical components (i.e., solenoids, sensors) will remain relatively constant.

Cost decreases through the use of fiberglass tanks for containing liquid heat energy storage media will be significant for smaller (i.e., residential) systems. Considerable economies of scale already exist for larger tanks. Table II presents expected cost impacts for these changes.

Installation and Design

These are the labor-intensive operations required to physically apply the SHACOB system to particular structures. Since each application is unique, it is currently necessary to perform analytical tasks to design a cost-effective SHACOB system. Cost for a design contractor or for design efforts by the installation contractor is currently incurred in commercially installed systems. Development of standardized design methodology, manuals, and design charts may result in as much as a 10 percent reduction in total system cost due to reduction in the need for highly specialized skills in the design of these systems.

Standardization of collector sizes and fittings or internal manifolding in the collectors might each reduce

Table I

Potential Cost Reductions for
SHACOB Collectors (Flat Plate)

PROCESS OR TECHNIQUE TO ENABLE COST REDUCTION	IMPACT ON COST
AUTOMATION OF PRODUCTION FACILITIES	$-4.00\$/F^2$
VAPOR DEPOSITION OF SELECTIVE SURFACES	$-0.20\$/F^2$
MATERIALS SUBSTITUTION GLAZING	$-0.35\$/F^2$

COLLECTORS (EVACUATED TUBE)

AUTOMATION OF PRODUCTION FACILITIES	$-14.00\$/F^2$

Table II

Potential Cost Reductions for
SHACOB Controls and Storage

PROCESS OR TECHNIQUE TO ENABLE COST REDUCTIONS	IMPACT ON COST $\$/Foot^2$ Collector
CONTROLS	
MASS PRODUCTION OF CONTROL CIRCUITRY	0.15
STORAGE	
USE OF FIBERGLASS TANKS	0.30

installation time by 20 percent. Combining these improvements could reduce aggregate installation time by a total of 30 percent.[3] Complete modularization of the SHACOB system may reduce installation costs by as much as 90 percent.

Combined improvements in design and installation costs of SHACOB systems might result in a saving of $7 per square foot of collector. This is summarized in Table III.

Cooling Devices

Cooling by means of solar energy collection generally employs absorption refrigeration or dessicant-type systems. The low-temperature type heat produced by solar collectors[4] has generally resulted in the need for derating of cooling devices.

It may be possible to decrease the cost of solar absorption equipment for a given cooling demand by increasing the fraction of rated capacity which may be achieved for hot liquid systems. Based on work by Butz,[5] it is estimated that adding a mechanical pump to reduce submergence and through improved heat exchanger design, useful capacity may be increased from 50 percent to 60 percent of machine rating. A net cost reduction of 12 percent is estimated to be possible (20 percent capacity increase with an 8 percent cost increase due to improved heat exchangers).

The use of a dessicant system sould remove the need for special heating and cooling subsystems. Although the dessicant device would cost about the same as an absorption device, the resulting system cost would be about 15 percent lower. Table IV summarizes these findings.

SHACOB COLLECTOR PERFORMANCE IMPROVEMENTS

Potential improvements in medium-temperature thermal collector performance lie in the following areas:

- Glazing transmittance may be improved through the use of non-reflective surfaces and glasses with lower iron content.
- Selective surface thermal parameters may approach the ideal α/ε of 0.95/0.05.[6] However, durability may be a problem.
- Heat losses may be further reduced through use of vacuums, honeycombs, convection suppressors and conduction insulators.

Table V lists the potential performance improvements.

Table III. Potential Cost Reductions for SHACOB System Installation and Design

PROCESS OR TECHNIQUE TO ENABLE COST REDUCTION	INSTALLATION	IMPACT ON MODULE COST
STANDARDIZATION OF COLLECTOR SIZES AND FITTINGS	-20% ⎫ -30%	-5% ⎫ -7.5%
INTERNAL MANIFOLDING AND COLLECTORS	-20% ⎭	-5% ⎭
MODULARIZATION OF SYSTEMS	-90% ⎱ -90%	-22.5% ⎱ 22.5%
DESIGN		
DEVELOPMENT OF DESIGN MANUALS AND CHARTS		-10%

Table IV. Potential Cost Reductions for SHACOB Solar Cooling Devices

PROCESS OR TECHNIQUE TO ENABLE COST REDUCTION	IMPACT ON COST MODULE	IMPACT ON TOTAL SYSTEM COST
ADD MECHANICAL PUMP TP REDUCE SUBMERGENCE AND REDESIGN HEAT EXCHANGERS TO ACHIEVE BETTER HEAT TRANSFER FROM HOT WATER TO THE SOLUTION IN THE GENERATOR IN ORDER TO INCREASE CAPACITY.	-12%	- 2%
USE OF DESSICANT SYSTEMS ALLOWING FOR NON-SEALED OPERATION, MULTIPLE USE (HEATING, COOL-ING, DELIGUIDIFICATION, LIQUIDIFICATION)		-15%

SHACOB SYSTEM COST SENSITIVITY

When SHACOB system size and performance are held
constant, the sensitivity of the various cost reduction
elements discussed previously may be evaluated in terms of
impact on life-cycle cost. Impacts for 10 to 20 percent
element cost reductions in residential applications are shown
in Table VI.

Of the cost reductions postulated earlier and for the
assumptions stated in Table VI, collector cost and cost of
systems design offer the potential for the greatest reduction
in system life-cycle cost.

COMBINED COST AND PERFORMANCE IMPROVEMENTS FOR SHACOB

The combined effects of both cost and performance
improvements are shown in Figures 1 and 2. A curve fit was
performed in order to show the cost of achieving various
efficiencies at three different fluid parameter values. The
generalized curves may be used to approximate cost/efficiency
relationships for specific applications and regions. This
implies working or average temperatures for the collectors and
specified ambient temperatures and insolation for the various
regions.

The fluid parameter is stated as:

$$\text{Fluid Parameter} = \frac{\text{Average Plate Temp.} - \text{Ambient Temp.}}{\text{Insolation}}$$

The 1976 curves are based upon manufacturers' data for
current products. The ultimate curves are based upon the
cost and performance potentials stated in preceding sections.

IMPACTS OF COST/PERFORMANCE IMPROVEMENTS FOR SHACOB

System initial costs and life-cycle costs may be
decreased significantly through cost and performance
improvements. System cost impacts may be viewed in three
different ways:

- Case I--If collector efficiency, collector area
  for a specific application and fluid parameter
  are held constant, the cost of collectors may
  decrease due to manufacturing improvements.
- Case II--If collector cost and fluid parameter
  are held constant but collector efficiency is
  increased to to advances in the technology, the
  amount of collector area required for a given

Table V

Performance Improvements for SHACOB Collectors

MODULE	PERFORMANCE MEASURE	PRESENTLY ASSUMED PERFORMANCE	MAXIMUM POTENTIAL	EFFICIENCY		
				FP*=.3	FP=.5	FP=.7
GLAZING	TRANSMITTANCE	.9	.95	.44	.27	.1
SELECTIVE SURFACE	ABSORPTIVITY	.9	.95	.48	.31	.14
	EMISSIVITY	.2	.05	.45	.28	.11
GLAZING SYSTEM	UPWARD HEAT LOSS U_L	.9 Btu/f^2/F°	.20	.275	.675	.625
COLLECTOR	BACK AND L	1.0	.25			
HOUSING	EDGE LOSSES	.1Btu/f^2/F°	.05			

*FP = FLUID PARAMETER = $\dfrac{\text{AVERAGE PLATE TEMPERATURE} - \text{AMBIENT TEMPERATURE}}{\text{INSOLATION}}$

Table VI

SHACOB System Sensitivity to Cost

MODULE	% COST REDUCTION IN MODULE	% COST REDUCTION IN TOTAL COST	% REDUCTION IN LCC**
COLLECTOR	10%	3%	3%
	20%	6%	6%
INSTALLATION	10%	1%	1%
	20%	2%	2%
DESIGN*	10%	3%	3%
	20%	5%	5%
CONTROLS	10%	<1%	<1%
	20%	<1%	<1%

* INCLUDES OVERHEAD, FEES, AND MARKUPS

** LIFE-CYCLE COST

ASSUMPTIONS: 5% inflation, 7% fuel escalation, $15/mmBtu delivered fuel cost,
33.3x10^6 Btu/yr load, 10% discount rate, 9%/30yr. mortgage, 20 year
analysis, 85% of load met by solar.

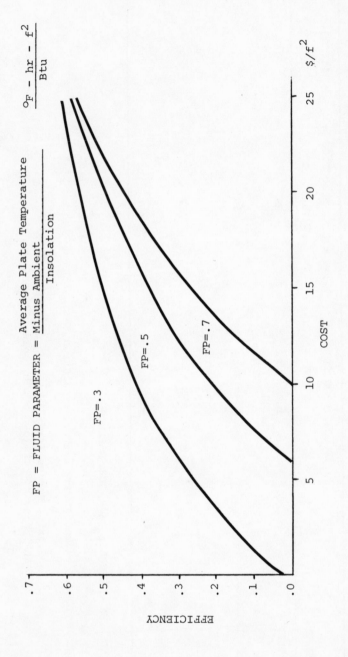

Figure 1.   Collectors: Cost vs Performance 1976.   Performance is plotted based on efficiency at various Fluid Parameter values from instantaneous efficiency curves.

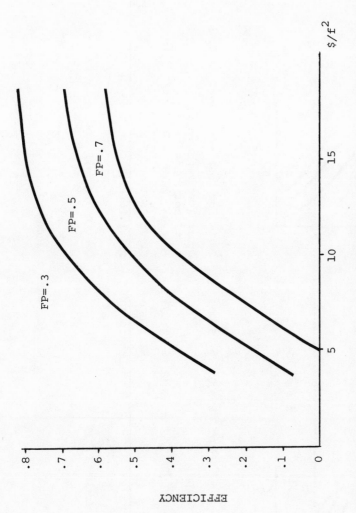

Figure 2.   Collectors: Cost vs Performance, Estimated Ultimate Potential.
Performance is plotted based on efficiency at various Fluid
Parameter values from instantaneous efficiency curves.

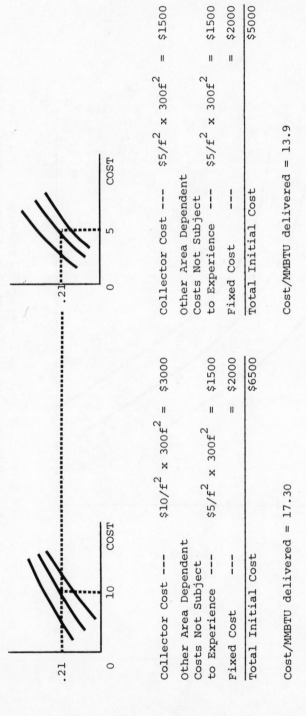

Collector Cost ---	$10/f^2 \times 300f^2 = \$3000$		Collector Cost ---	$\$5/f^2 \times 300f^2 = \$1500$
Other Area Dependent Costs Not Subject to Experience ---	$\$5/f^2 \times 300f^2 = \$1500$		Other Area Dependent Costs Not Subject to Experience ---	$\$5/f^2 \times 300f^2 = \$1500$
Fixed Cost ---	$= \$2000$		Fixed Cost ---	$= \$2000$
Total Initial Cost	$\$6500$		Total Initial Cost	$\$5000$
Cost/MMBTU delivered = 17.30			Cost/MMBTU delivered = 13.9	

Figure 3. Case I Example

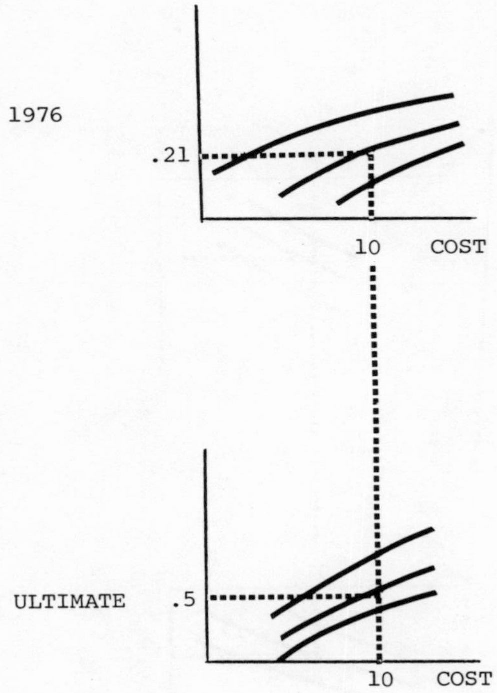

Collector Cost --- $10/f^2 \times 300f^2$ =$3000

Other Area Dependent
Costs Not Subject
to Experience --- $5/f^2 \times 300f^2$ =$1500

Fixed Cost --- =$2000

Total Initial Costs $6500

Cost/MMBTU Delivered 17.30

Collector Cost --- $10/f^2 \times 125f^2$ =$1250

Other Area Dependent
Costs Not Subject
to Experience --- $5/f^2 \times 125f^2$ =$ 625

Fixed Costs --- $2000

Total Initial Costs $3875

Cost/MMBTU Delivered 12.2

$$A_{ULTIMATE} = \frac{A_{1976}}{.5} \over .21$$

1976

.21

10 COST

ULTIMATE

.5

10 COST

Figure 4. Case II Example

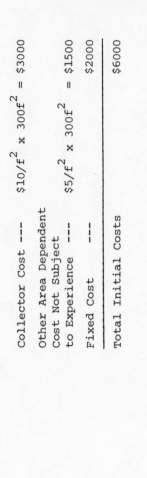

$$A_{ULTIMATE} = A_{1976}$$

Collector Cost ---	$10/f^2$ x $300f^2$ =	$3000
Other Area Dependent Cost Not Subject to Experience ---	$5/f^2$ x $300f^2$ =	$1500
Fixed Cost ---		$2000
Total Initial Costs		$6000

Cost/MMBTU Delivered (1976)	17.3
Cost/MMBTU Delivered (Ultimate)	11.4

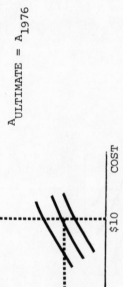

Figure 5. Case III Example

application will be reduced, thereby reducing system cost.

● Case III--If collector cost and area and fluid parameter are held constant, thereby holding system cost constant, collector efficiency improvements will increase system heat output without an increase in system cost.

Figures 3, 4 and 5 show examples of cost analysis for the three cases cited above. In the first case, a 50% reduction in collector costs (holding efficienty and array size constant) results in only a 23% reduction in system cost. In the second case, because of technological advances, the average efficiency of the collector that you may purchase for $10/f^2 increases greatly thereby decreasing the area of collector required to provide approximately the same amount of energy. Thus, in this case not only is the total collector array cost decreased because of the smaller size, the other area dependent costs are also decreased resulting in a 40% reduction of system costs. In case three, the impacts of technological performance improvements and cost decreases are both taken into account by allowing more collector array which performs at a higher efficiency to be installed at a total system cost no greater than what is being payed today. In this case, a much larger portion of the load is met by solar thus the cost of delivered energy is greatly decreased even though the total system cost remains the same.

Reduction in the cost of energy from SHACOB systems will yield increasing market penetration.

Table VII

Impacts on Market Penetration of
Improved Cost/Performance in
SHACOB Systems
(Percent of Specific Market)

	1980	1985	2000
PRESENT COST/PERFORMANCE	3.3	9.5	20.2
CASE I	4.5	13.3	34.8
CASE II	5.9	17.5	48.6
CASE III	3.6	10.5	24.1

Table VII shows the impact on market penetration for each of the three cases of cost/performance improvements. The specific example for which the table is derived includes new one- and two-family residences in California which would otherwise use electric resistance heating and hot water systems.

REFERENCES

(1)   Various concentrating collectors are alternatives but are not so widely employed.

(2)   Latent heat storage (i.e., latent heat of fusion or solid-liquid phase change) systems are being developed but are, as yet, very expensive.

(3)   Combined effects, because of diminishing returns, would not be the sum of the individual effects.

(4)   For flat plate and evacuated-tube collectors. Concentrating collectors produce higher temperatures but at a significantly higher cost.

(5)   Butz, L.W., "Use of Solar Energy for Residential Heating and Cooling," M.S. Thesis, University of Wisconsin, 1973.

(6)   S. Klein, J. Duffy and W. Beckman, Solar Heating Design By The F-CHART Method, John Wiler and Sons, Inc., New York, New York 10016, 1977.

Optimal Design of Seasonal Storage for 100% Solar Space Heating in Buildings

9

Ronald O. Mueller, Joseph G. Asbury, Joseph V. Caruso,
Donald W. Connor, and Robert F. Giese

This paper presents study findings on the thermal performance, subsystem sizing and relative economics of "seasonal" solar space heating systems (systems containing large
storage capacities on the order of one or several month's
supply). The principal rationale of seasonal storage is to
permit solar collection outside of the heating season, thereby improving the utilization of the total available solar
input, and reducing collector area requirements. Under the
assumption of a fully-mixed sensible storage medium, we develop a concise system model, one that integrates over the
sub-daily fluctuations in insolation and load. Under average
year weather conditions, steady state solutions are derived
for the specific class of system designs in which solar
energy meets the entire heating load. For these 100% solar
systems, approximate analytic expressions are derived that
relate the sizing and design requirements of the collector
and storage subsystems. Based on these performance findings,
the total costs of seasonal solar systems are compared
against the costs of both "diurnal" storage solar systems and
conventional heating systems, with the results presented in
terms of seasonal storage break-even cost estimates. As a
final step in the analysis, we consider the alternatives
available for meeting a "worst-case" heating season and make
rough comparisons of the cost tradeoffs of either oversizing
the seasonal solar system or incorporating a small backup
heating device (1).

Introduction

In many applications of solar energy — space heating
and cooling of buildings are two examples — load requirements are non-zero only over a fraction of the year. In

meeting these periodic loads with "diurnal" solar systems (systems containing storage capacities on the order of a day's supply) the solar radiation incident on the collector field during the off-season remains largely unused because of the lack of contiguous demand. This can represent a substantial opportunity loss in energy collection. For example, in winter space heating of buildings the fraction of solar energy that falls outside of the main winter months may reach as high as 60%. Because of the capital-intensive and fuel-unintensive nature of solar technologies, the cost of delivered solar energy will generally be higher for periodic loads than in applications where the load is relatively constant over the year. Recently, a number of solar system designs have been proposed, some built, that incorporate storage capacities on the order of one or several months' supply (1a). The rationale for these "seasonal" solar schemes is to permit solar collection over a much greater fraction of the year, with energy stored for periods extending over several months or longer before actually being used to meet load.

By improving utilization of the available solar input over the full year, a seasonal system will reduce collector area requirements over that of a comparable diurnal system while providing the same total energy to load. The resultant savings in collector area costs represent a major economic benefit of seasonal storage. A second benefit is to lessen, or eliminate altogether, requirements for a backup energy supply and the associated problems of load management (2). By providing a reliable, long-term buffer between short-term stochastic variations in the solar input and the load, seasonal storage permits design of solar systems that supply all, or nearly all, of the total energy requirements. Against these benefits must be weighed the added capital cost of the storage unit.

In this paper we present an analysis of a seasonal solar system design consisting of conventional flat plate collectors and a sensible heat storage medium. A concise system model is developed under the assumption of a fully-mixed, uniform temperature, storage medium that permits efficient simulation of long-term (multi-day) system thermal performance over the course of the year. The approach explicitly neglects the effects on storage temperature of short-term (sub-daily) fluctuations in insolation and load, effects that will be extremely small for seasonal solar systems. Although perhaps not adequate as a detailed design tool, this approach is useful for examining the major design tradeoffs of concern in this paper. The specific application considered is winter space heating of buildings, although the approach adopted will be useful for other periodic loads as well.

The analysis proceeds through two stages. First, we solve for the thermal performance of seasonal solar systems that are designed to supply 100% of load without any backup, under "reference year" monthly normal ground temperature and insolation conditions. The systems are matched to the load requirements of a 150 m^2, well-insulated, detached single family dwelling unit. Although not considered here, a similar approach could be applied to partial seasonal systems, supplying less than full load requirements. For the class of 100% solar systems, approximate analytic expressions are derived that relate sizing and design requirements of the collector to the storage component. Based on the performance findings we estimate unit break-even costs of seasonal storage by comparing the capital and fuel costs of conventional space heating technologies against those of seasonal solar systems. At costs below the break-even estimates, the seasonal solar system has an economic advantage over the conventional system. We also make cost comparisons between seasonal and diurnal solar systems.

Seasonal solar systems designed to meet load under average "reference year" weather conditions will fall short during the later stages of more severe winters. To avoid this shortfall, either the seasonal solar system can be oversized or a small backup heating system attached. As a second step in the analysis we have made a rough comparison of the cost tradeoffs between these alternatives, by examining statistical variations in winter season conditions over the past several decades.

The four northerly sites for which detailed results are presented here are: Caribou, Maine; Madison, Wisconsin; Boston, Massachusetts; and Sterling, Virginia. Provided the storage vessel is extremely well insulated, we find substantial performance gains for the seasonal over the diurnal system, with the annual fraction of useful solar energy delivered to load greater by as much as a factor of two under reference year weather conditions. For the range of collector costs and conventional system costs considered in the paper, the corresponding storage break-even costs range up to a high value of about 15¢/gallon ($40/m^3) for extremely well insulated tanks.

Although the storage break-even costs are dependent on the costs assumed for collectors and for the conventional technologies, the qualitatively low range of estimates is symptomatic of a seasonal storage system. In simplest terms, the value of each increment of storage capacity is set by the cost difference between the aggregate energy input to storage (divided by storage efficiency) and a comparable amount of

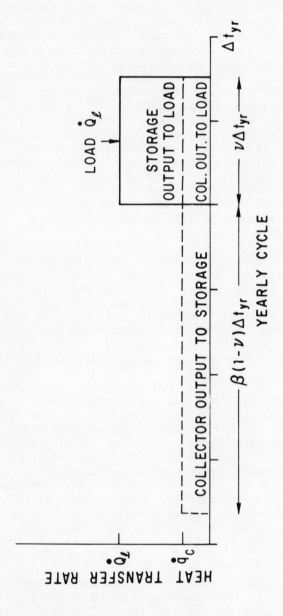

Figure 1. Heat Transfer from Collector to Storage, Storage to Load, and Collector to Load over the Yearly Cycle for the Simplified "Square-Wave" Load $\dot{Q}_\ell$

competing conventional fuel. For a seasonal system, thermal
energy is cycled through the device only about once per year,
leading to a relatively low annual energy throughput and
hence to a low storage break-even cost. By contrast, in
diurnal solar applications, energy is cycled through the
storage device numerous times over the heating season, increasing proportionately the total energy throughput and the
storage break-even cost.

The remainder of the paper is organized as follows: In
Section 2 we present a simplified model for estimating storage break-even costs of seasonal solar systems compared
against diurnal solar systems as well as against conventional
heating devices. In Section 3 we present a model of a seasonal solar system, that is solved in Section 4 for long-term
system performance using a Fourier series approach. Using the
sizing relations developed in Section 4, we make detailed
cost comparisons of seasonal solar systems versus conventional systems in Section 5, developing estimates of storage
break-even costs as a function of the major system and cost
parameters. In this final section we also consider the cost
tradeoffs associated with the design of seasonal systems
capable of meeting load under "worst-case" winter conditions.
A list of nomenclature used in the paper follows Section 5.

2. Storage Break-Even Costs:
Simple Estimates

The analysis presented in this section serves to illustrate the role of seasonal storage, while offering simple
estimates of storage break-even costs under a number of simplifying assumptions. Figure 1 depicts the application of a
seasonal solar system in supplying energy to meet the
"square-wave" heating load $\dot{Q}_\ell$, which is assumed constant over
the fraction ν of the year. By permitting collection and
storage of solar energy during the $(1 - \nu)$ "off-season" fraction of the year, a seasonal solar system requires less
collector area to meet load than a comparable diurnal solar
system. In calculating the reduction in collector area, we
make the further simplifying assumption that daily insolation
is constant over the year and not subject to random outages.

If the storage tank is assumed perfectly insulated, and
temperature-dependent collector losses are ignored, the
reduction in collector area is given simply by

$$\frac{A_c' - A_c}{A_c'} = 1 - \nu .$$

(1)

A_c and A_c' are the area requirements of the seasonal and diurnal solar systems, respectively, with both devices designed to meet the full load. If the daily average collection rate per unit area is denoted $\dot{q}_c$, then the area A_c' of the diurnal system is just $\dot{Q}_\ell/\dot{q}_c$. Losses from the seasonal storage tank during the off-season will tend to increase collector area requirements A_c over that specified in Eq. 1. If we denote by β the fraction of off-season solar output that is actually delivered to load, the reduction in area requirements can be represented by

$$\frac{A_c' - A_c}{A_c'} = \frac{\beta(1 - \nu)}{\nu + \beta(1 - \nu)} \; . \tag{2}$$

The net savings in collector related capital costs is then given by

$$c_c(1 - A_c/A_c')\dot{Q}_\ell/\dot{q}_c \; , \tag{3}$$

where c_c represents unit collector capital costs.

Against the net cost savings in the collector unit must be weighed the added capital cost of the storage unit. Assuming water is the storage medium the volume V_s of storage required to maximize the reduction in collector area is given by

$$V_s = \frac{\nu\Delta t_{yr}\dot{Q}_\ell}{\beta\Delta T_s\rho c_p} (1 - A_c/A_c') , \tag{4}$$

where ρ is the density and c_p the specific heat of water. Δt_{yr} represents a year and ΔT_s is the yearly variation in storage temperature, with the storage tank assumed fully charged at the onset of the load. In Eq. 4 the β^{-1} factor accounts for the oversizing of storage needed to offset thermal losses, while the factor $(1 - A_c/A_c')$ accounts for the fraction of load met directly by collector output. Denoting the cost per unit storage volume by c_s, the cost of the storage unit is then c_sV_s. Returns to scale on the unit cost of storage are neglected.

Equating storage costs to collector savings, and solving for c_s yields

$$\tilde{c}_s = c_c(A_c' - A_c)V_s^{-1} \; . \tag{5}$$

$\tilde{c}_s$ represents the unit break-even cost of storage for which

the seasonal and diurnal solar systems have equal total
costs. (In this comparison we have ignored the relatively
small cost of storage for the diurnal system.)

Analogously, we can compare the relative economics of a
seasonal solar system against a conventional heating system.
The unit capital and fuel cost components of the conventional
system are denoted c_{conv} and c_F, respectively, with the total
annualized cost then represented by

$$c_{conv}\dot{Q}_\ell i + \frac{Lc_F \nu \Delta t_{yr} \dot{Q}_\ell}{\eta} , \tag{6}$$

where i is the capital charge rate, L the fuel levelization factor over the life of the device and η the fuel conversion efficiency of the device (3). Equating the total
annualized costs of the seasonal solar and conventional
heating technologies and solving for c_S yields

$$\tilde{c}_s' = \frac{1}{V_s} \left[c_{conv}\dot{Q}_\ell + \frac{Lc_F \nu \Delta t_{yr} \dot{Q}_\ell}{\eta i} - c_c A_c \right] . \tag{7}$$

$\tilde{c}_s'$ represents the break-even cost of seasonal storage relative to the conventional heating technology.

As an example consider a load of six months' duration,
$\nu = 1/2$, and an off-season storage efficiency of 70%, $\beta = 0.7$.
From Eq. 2 the ratio of collector areas A_c/A_c' is computed to
be about 0.6. The cost comparison of seasonal versus diurnal
storage can then be made using Eq. 5. Taking typical para-
meter values, $\dot{q}_c = 360$ kJ/hr $- m^2$, $\Delta T_s = 40°C$ and $c_c = \$100/m^2$,
Eq. 5 yields an estimate of $\tilde{c}_s$ of about 2.7¢/gal. ($\$7.2/m^3$).
Analogously Eq. 7 can be used to make a cost comparison of
seasonal storage versus conventional heating. For i = 0.1
and L = 1.3, and for the following conventional technology
parameter values, $c_{conv} = \$50/kW$, $\eta = 0.55$, $c_F = \$3/10^6 kJ$
(1¢/kWh), the value of $\tilde{c}_s'$ is calculated to be about 11¢/gal
($\$29/m^2$). At zero collector-related costs $\tilde{c}_s'$ is 20¢/gal
($\$53/m^2$).

3. Model Description

A schematic of the seasonal solar system design studied
in this report is given in Fig. 2. The system is assumed
hydronic, with the storage medium fully mixed (isothermal)
at temperature T_s. The collector and load connections are
attached separately to the storage module, requiring solar
energy to pass from collector to storage and then to load (4).

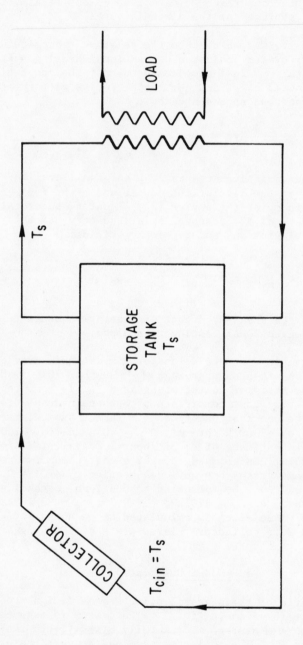

Figure 2. Schematic of a Seasonal Solar System Containing a Fully Mixed (Isothermal) Storage Medium

Representing the instantaneous thermal power output from the collector unit by the Hottel Whillier equation, we can integrate this over the course of a day's solar charging period to obtain the total energy collected

$$\Delta Q_c = A_c F_R \left[(\tau\alpha) H_T - f u_c (<T_{cin}>_c - <T_A>_c) \right] \Delta t. \tag{8}$$

The collector specific parameters are: u_c, collector heat loss coefficient; $(\tau\alpha)$, transmittance absorptance product; F_R, collector heat removal factor; and A_c collector area. H_T is the average over the day, $\Delta t \equiv 24$ hours, of solar radiation incident on the inclined collector surface. Thermal losses from the collector unit occur only during the fraction f of the day that the collectors operate. $<T_{cin}>_c$ and $<T_A>_c$ represent the average of collector inlet temperature and ambient temperature during the collection period. In applying Eq. 8 to a seasonal solar system, the analysis is simplified considerably by approximating $<T_{cin}>_c$ by the average daily storage temperature T_S. (Note, we generally adopt the convention that temperature variables refer to daily average values, unless specified otherwise.) For a seasonal system this represents a reasonable approximation since the total storage heat capacity is sized well above daily collector and load heat requirements, and where as a result, changes in storage temperature over a day are small, generally less than a few degrees Celsius.

To simplify notation, we recast Eq. 8 by combining (for a specific collector type) insolation and ambient temperature into a single variable, an "effective" collector stagnation temperature,

$$T_c = U_c^{-1} (\tau\alpha) H_T + <T_A>_c , \tag{9}$$

where $U_c \equiv f u_c$. T_c is formally equal to the value of collector fluid inlet temperature for which heat collection goes to zero. Using Eq. 9, and setting $<T_{cin}>_c$ equal to T_S, Eq. 8 becomes for $F_R = 1$.

$$\Delta Q_c = A_c U_c (T_c - T_S) \Delta t, \tag{10}$$

applicable for $T_S < T_C$. In the fall and early winter seasons lower ambient temperatures and insolation levels cause T_C to decrease, and T_S may actually exceed T_C for brief periods until load requirements lower storage temperatures sufficiently. During this period collectors are assumed not to operate, and ΔQ_c is set equal to zero.

The total daily load requirement during the heating
season is represented exogenously by

$$\Delta Q_\ell = A_\ell U_\ell \tilde{T}_\ell \Delta t, \tag{11}$$

valid for $\tilde{T}_\ell > 0$; ΔQ_ℓ being set to zero for $\tilde{T}_\ell \leq 0$. $\tilde{T}_\ell$ is an
effective daily average load temperature difference, A_ℓ is
the total area of the building shell, and U_ℓ the average heat
loss coefficient per unit area of the structure. Although,
in principle, Eq. 11 can represent all or only a fraction of
a building's daily load (with the remainder made up by a
conventional heating system), the analysis below assumes ΔQ_ℓ
to be the full load requirement. Within the simple degree-
day approach adopted here, $\tilde{T}_\ell$ reduces to $T_r - T_A$ with T_r the
constant room temperature setting and T_A the daily average
ambient temperature.

We bypass the complication of defining a load heat-
exchanger equation, by assuming that the seasonal solar system is always capable of meeting the load requirements
specified by Eq. 11. Implicit in this approach are the
assumptions that: (1) the load heat-exchanger is sized
adequately to meet design heating conditions and (2) throughout the winter season storage temperatures remain above a
minimum value adequate for space heating purposes. The solutions presented in the next section are explicitly required
to satisfy the latter assumption.

Besides thermal output to load, inevitable tank losses
can represent a substantial (undesirable) thermal drain on
the storage unit, occurring year-round. We represent these
losses by

$$\Delta Q_t = A_t U_t \tilde{T}_t \Delta t, \tag{12}$$

with $\tilde{T}_t$ an effective daily average temperature difference
between the storage tank and its surrounding environment. U_t
is the average heat loss coefficient for the storage tank and
A_t is the tank surface area, with A_t roughly proportional to
the 2/3 power of tank volume V_s. Although unlikely to occur
in a space heating application, the storage medium can gain
energy from its surroundings provided $\tilde{T}_t$ is negative. In the
present analysis we assume the storage vessel is buried
underground, and approximate $\tilde{T}_t$ by the simple temperature
difference $T_s - T_g$, with the ground temperature T_g constant
and independent of T_s (5). The corresponding value of U_t in
Eq. 12 is taken to include the composite thermal resistance
of tank insulation and surrounding earth.

Balancing the net of daily thermal inputs and outputs to
the storage unit, to the change in its internal sensible
energy leads to an equation defining the daily change in
tank temperature

$$V_s \rho c_p \Delta T_s / \Delta t = (\Delta Q_c - \Delta Q_\ell - \Delta Q_t)/\Delta t, \tag{13}$$

with ρ the storage mass density and c_p its specific heat.
With substitution of Eqs. 10-12, and rearrangement of terms,
Eq. 13 can be written

$$\frac{\Delta T_s}{\Delta t} + (\lambda_c + \lambda_t)T_s = \lambda_t T_g + \lambda_c T_c - \lambda_\ell \tilde{T}_\ell, \tag{14}$$

where λ's are component time constants defined by:
$\lambda_i = A_i U_i / V_s \rho c_p$. Eq. 14 forms the basic defining equation of
system performance that can be applied to both 100% and par-
tial seasonal solar systems. Given knowledge of the temperature variables T_c and $\tilde{T}_\ell$ over the year, and an initial value
for $T_s(t_o)$, the solution for T_s can be obtained directly
using a numerical finite difference approach. An alternate
approach, adopted in the following section, is to develop a
Fourier-series representation for T_s that hinges upon the
roughly periodic behavior of the driving temperatures T_c and
$\tilde{T}_\ell$.

4. System Analysis

This section presents an analysis of system performance,
via Eq. 14, under the assumptions that: (1) the seasonal
solar system faces reference year, monthly normal, weather
conditions, (2) has achieved steady state operating conditions, and (3) is capable of meeting the full heating load
without auxiliary backup (6,7). Table 1 gives the monthly
normal values for heating degree days per day, which are
equivalent to $\tilde{T}_\ell$, and daily global insolation on a tilted
surface, $H_T \Delta t$, for the four sites considered in the report:
Caribou, Maine; Madison, Wisconsin; Boston, Massachusetts;
and Sterling, Virginia (8,9).

Although less than adequate for analysis of diurnal
systems, the use of reference year monthly normal weather
data provides a valid benchmark for seasonal systems. Provided storage capacities are large, seasonal systems will
integrate out short-term stochastic fluctuations in insolation and load, with system performance dependent primarily
on long-run (weekly, monthly) average values. Below, we
determine sizing requirements of seasonal systems that
exactly meet load under reference year weather conditions.

Table 1. Long Term Monthly Normals of Heating Degree Days and Isolation[a]

	Location	Monthly Normals												Annual Heating Degree Days
		Oct.	Nov.	Dec.	Jan.	Feb.	Mar.	Apr.	May	June	July	Aug.	Sept.	
Degree Days per Day[b]: $\tilde{T}_\ell$(°F)	Sterling	9.4	20.3	31.0	32.9	30.9	23.2	11.9	4.2	0.2	0.0	0.0	1.4	5010
	Boston	9.7	19.8	32.0	35.8	34.3	26.9	16.4	7.0	0.9	0.0	0.3	2.5	5620
	Madison	15.3	30.3	43.1	48.2	44.3	34.8	19.7	9.6	2.4	0.5	1.3	5.8	7730
	Caribou	21.2	33.6	48.9	54.3	51.6	41.4	28.3	15.3	5.7	2.7	3.9	10.9	9630
														TILT
Total Daily Insolation[c]: $H_T \Delta t$ (10⁴ kJ/m²)	Sterling	1.65	1.40	1.23	1.23	1.48	1.71	1.79	1.81	1.90	1.81	1.79	1.79	45°
	Boston	1.40	1.00	0.99	1.07	1.26	1.46	1.48	1.67	1.63	1.67	1.59	1.57	53°
	Madison	1.62	1.08	1.21	1.32	1.47	1.76	1.61	1.66	1.78	1.84	1.78	1.84	53°
	Caribou	1.34	0.85	1.12	1.32	1.80	2.15	1.74	1.75	1.62	1.77	1.78	1.66	53°

[a] See Ref. 8 for specific citations to data sources.
[b] Base 65°F.
[c] Insolation on tilted surface.

Assuming the exogenous temperatures T_c and $\tilde{T}_\ell$ are periodic over the year, the corresponding periodic (steady-state) solution for T_s can be constructed directly by substituting in Eq. 14 Fourier-series representations for all temperature variables of the form

$$T_i = T_i^\circ + \sum_{n=1}^{N}\left[T_{i1}^n \sin(n\omega t) + T_{i2}^n \cos(n\omega t)\right], \tag{15}$$

with ω the fundamental angular frequency (2π/year). With October 1 taken as the start of the yearly cycle (just before onset of the heating season), Table 2 gives the leading Fourier-series coefficients for $\tilde{T}_\ell$ and H_T, evaluated for each site using the monthly normal data in Table 1. Using Eq. 9 the corresponding Fourier-series coefficients for T_c can be calculated for specific values of the collector parameters ($\tau\alpha$) and U_c (10).

Approximating the finite difference term $\Delta T_s/\Delta t$ by its limiting derivative dT_s/dt, Eq. 14 can be reduced to a set of algebraic equations by equating the sum of coefficients of each Fourier-series term to zero. This yields a single equation for the coefficient of the time independent term T_s°

$$T_s^\circ = (\lambda_c + \lambda_t)^{-1}(\lambda_t T_g + \lambda_c T_c^\circ - \lambda_\ell \tilde{T}_\ell^\circ), \tag{16}$$

and a set of coupled equations for the nth order coefficients $\{T_{s1}^n, T_{s2}^n\}$

$$\begin{aligned}
(\lambda_c + \lambda_t)T_{s1}^n - n\omega T_{s2}^n &= \lambda_c T_{c1}^n - \lambda_\ell \tilde{T}_{\ell 1}^n \\
n\omega T_{s1}^n + (\lambda_c + \lambda_t)T_{s2}^n &= \lambda_c T_{c2}^n - \lambda_\ell \tilde{T}_{\ell 2}^n.
\end{aligned} \tag{17}$$

Equations 16 and 17 provide explicit solutions for the tank temperature coefficients. An implicit assumption in their derivation is that the parameters U_c, U_ℓ and U_t are strictly constants independent of time. While adequate for our present treatment, the values of these parameters will in fact vary over the course of the year. U_c, for example, will be larger during summer than winter by about 30% because of the greater number of daylight hours in summer that causes the fraction f to increase. Although not detailed here, a more precise treatment would require Fourier-series representations for these parameters comparable to those used for the temperature variables. Although still reducible to a set of algebraic equations, the resulting solutions would be much

Table 2. Leading Fourier Series Coefficients for $\tilde{T}_\ell$ and H_T [a,b]

Location	$\tilde{T}_\ell^0$	$\tilde{T}_{\ell 1}^1$	$\tilde{T}_{\ell 2}^1$	$\tilde{T}_{\ell 1}^2$	$\tilde{T}_{\ell 2}^2$	$\tilde{T}_{\ell 1}^3$	$\tilde{T}_{\ell 2}^3$	$\tilde{T}_{\ell 1}^4$	$\tilde{T}_{\ell 2}^4$
$\tilde{T}_\ell$ (°F)									
Sterling	13.8	16.8	-5.8	-1.5	-2.7	0.3	-0.9	-0.22	-0.07
Boston	15.5	17.2	-7.8	-1.9	-2.0	-0.3	-0.6	-0.06	-0.11
Madison	21.3	23.1	-8.4	-1.7	-2.8	0.	-0.5	-0.33	-0.09
Caribou	26.5	24.3	-10.0	-1.6	-1.5	-0.7	-0.01	-0.05	0.36

Location	H^0	H_1^1	H_2^1	H_1^2	H_2^2	H_1^3	H_2^3	H_1^4	H_2^4
H_T (kJ/hr-m^2)									
Sterling	680	-125.	-11.	-3.	47.	15.	6.	9.	6.
Boston	585	-137.	-15.	-23.	33.	3.	21.	2.	8.
Madison	659	-119.	-8.	-56.	42.	6.	20.	12.	3.
Caribou	657	-108.	-109.	-110.	65.	9.	5.	21.	12.

[a]Calculated from monthly normal data presented in Table 1.

[b]Start of yearly cycle is October 1.

more complex. It appears that beyond the simplest sinusoidal expansion discussed below it is more efficient to directly solve Eq. 14 using a finite difference approach.

Over the year, the tank temperature T_s assumes its maximum and minimum values $\{T_s^{max}, T_s^{min}\}$ at the times $\{t_{max}, t_{min}\}$ given as solutions to the secular equation

$$dT_s/dt = \sum_{n=1}^{N} (n\omega) \left[T_{s1}^n \cos(n\omega t) - T_{s2}^n \sin(n\omega t) \right] = 0. \quad (18)$$

The yearly maximum and minimum values of tank temperature, given explicitly by

$$T_s^{max} = T_s(t_{max})$$

$$T_s^{min} = T_s(t_{min}), \quad (19)$$

play a key role in setting collector area and storage volume requirements. The values of these two temperature parameters are generally restricted by physical constraints within the solar system, with T_s^{min} required to be high enough to provide adequate heat transfer to load and T_s^{max} low enough to prevent structural damage to the storage unit. For specified values of $\{T_s^{max}, T_s^{min}\}$, and with the corresponding values of $\{t_{max}, t_{min}\}$ calculated from Eq. 18, Eqs. 19 can, as we exhibit below, be inverted to provide direct sizing estimates for collector area and storage volume in terms of the physical parameters that define the system and the exogenous temperature variables.

For the remainder of this section, we specialize results to the case where temperature variables are assumed to have a simple sinusoidal behavior at the fundamental frequency $\omega = 2\pi/year$ (that is, all Fourier-series are terminated beyond n=1). The particular merit of this simple case is that one can obtain from Eqs. 19 analytic relations for collector area and storage volume requirements. In general, the effects of higher order harmonics have been found to be relatively minor, with the results of the "sinusoidal" case adequate for the "first-cut" feasibility analysis presented here.

The solutions to Eq. 18 for $\{t_{max}, t_{min}\}$ reduce in this case to

$$\tan \omega t_{max} = T^1_{s1}/T^1_{s2}$$

$$t_{min} = t_{max} + 1/2 \text{ year},\tag{20}$$

with the times equally spaced at half-year intervals. The corresponding values for $\{T^{max}_s, T^{min}_s\}$ are

$$T^{max}_s = T^{\circ}_s + \left[T^1_{s1} \sin (\omega t_{max}) + T^1_{s2} \cos (\omega t_{max})\right]$$

$$T^{min}_s = T^{\circ}_s - \left[T^1_{s1} \sin (\omega t_{max}) + T^1_{s2} \cos (\omega t_{max})\right].\tag{21}$$

With substitution of Eqs. 16 and 17 for T°_s, T^1_{s1}, T^1_{s2}, and rearrangement, Eqs. 21 can be rewritten in terms of the average yearly storage temperature $\overline{T}_s$ and yearly storage temperature fluctuation ΔT_s

$$\overline{T}_s \equiv \tfrac{1}{2}(T^{max}_s + T^{min}_s) = T^{\circ}_s = (\lambda_c + \lambda_t)^{-1}(\lambda_t T_g + \lambda_c T^{\circ}_c - \lambda_{\ell}\tilde{T}^{\circ}_{\ell})\tag{22}$$

$$\Delta T^2_s \equiv (T^{max}_s - T^{min}_s)^2 = \frac{4\left[(\lambda_c T^1_{c1} - \lambda_{\ell}\tilde{T}^1_{\ell 1})^2 + (\lambda_c T^1_{c2} - \lambda_{\ell}\tilde{T}^1_{\ell 2})^2\right]}{\left[(\lambda_c + \lambda_t)^2 + \omega^2\right]}.\tag{23}$$

These equations can be inverted to provide the following explicit relations for collector area and storage volume requirements

$$A_c U_c = (T^{\circ}_c - \overline{T}_s)^{-1}\left[A_{\ell}U_{\ell}\tilde{T}^{\circ}_{\ell} + A_t U_t (\overline{T}_s - T_g)\right]\tag{24}$$

$$V_s/A_c = U_c/\omega\rho c_p \left[4(T^1_{c1} - \gamma\tilde{T}^1_{\ell 1})^2 + 4(T^1_{c2} - \gamma\tilde{T}^1_{\ell 2})^2\right.$$

$$\left. - \Delta T^2_s (1 + \delta)^2\right]^{\frac{1}{2}}\Delta T^{-1}_s,\tag{25}$$

where γ, δ are the ratios $A_{\ell}U_{\ell}/A_c U_c$ and $A_t U_t/A_c U_c$, respectively.

Eqs. 24 and 25 are a coupled set of equations that can be solved simultaneously to provide unique solutions for A_c and V_s/A_c. Their values represent the minimum collector area and storage volume requirements adequate to just meet load, under the specified constraints on the minimum and maximum yearly tank temperature. Collector area requirements depend only on the yearly average values of the exogenous temperature variables, while storage volume requirements depend upon the yearly fluctuations in temperatures. The yearly average storage temperature $\overline{T}_S$ affects collector area requirements directly through its effect on collection efficiency and indirectly through its effect upon storage tank losses, with A_c reduced as $\overline{T}_S$ is lowered. One way to reduce $\overline{T}_S$ without adversely affecting volume requirements is to lower T_S^{min}, and in what follows we assume that T_S^{min} is always set at the minimum value consistent with heat transfer to load requirements. To reduce $\overline{T}_S$ and hence A_c by lowering T_S^{max} also decreases ΔT_S and has the simultaneous effect of increasing storage volume V_S.

Within the sinusoidal approximation an explicit expression can be derived for the annual system efficiency, ε, defined here as the percent of total solar energy actually transferred to load over the year

$$\varepsilon = \frac{\int_{yr} \dot{Q}_\ell dt}{A_c \int_{yr} H_T dt} = (\tau\alpha) \cdot \frac{A_\ell U_\ell}{A_c U_c} \cdot \frac{\tilde{T}_\ell^\circ}{(T_c^\circ - T_A^\circ)} \cdot \qquad (26)$$

Figures 3A-D and 4A-D display graphically the behavior of the solutions to Eqs. 24-26 at all four sites, giving A_c, V_S, V_S/A_C and ε as functions of ΔT_S for different levels of tank insulation. The specific values of the load and system parameters held constant in these calculations are listed in Table 3. One of the principal results apparent from these graphs is the pronounced effect storage tank losses have on overall system efficiency. In Figs. 3A and 4A, the difference in area requirements between a specific R-value curve and the corresponding R = ∞ curve represents the excess area requirements needed solely to replenish storage losses over the year. The effect can be substantial, approaching 50% additional collector area at higher values of ΔT_S. In order to keep storage losses (and hence additional collector area) at a manageable level, say in the range of 10-20% of the total energy collected over the year, storage vessels with extremely high insulation properties are required with R values above 80 (Btu/hr - °F - ft^2)$^{-1}$.

As tank insulation improves, both collector area and storage volume requirements decrease, with the system producing a higher overall yearly efficiency. Higher tank insulation levels also decrease the ratio of storage volume to collector area, by shifting the optimum design to include more storage volume and less collector area. At the higher values of ΔT_S the ratio V_S/A_C is in the relatively low range

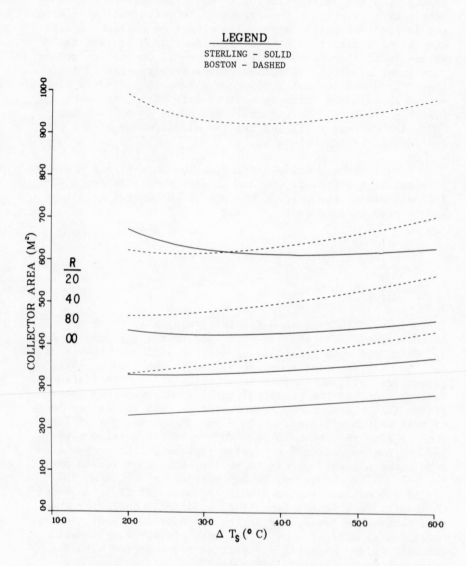

Fig. 3A-B. Collector Area and Storage Volume Requirements
in Sterling and Boston as a Function of Yearly
Variation in Storage Temperature ΔT_S, calculated

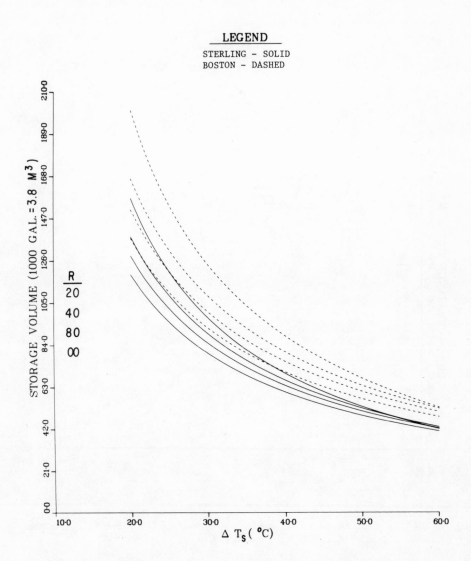

from Eqs. 24-25. The four curves, from top to
bottom respectively, correspond to different
levels of tank insulation, R = 20, 40, 80, ∞.

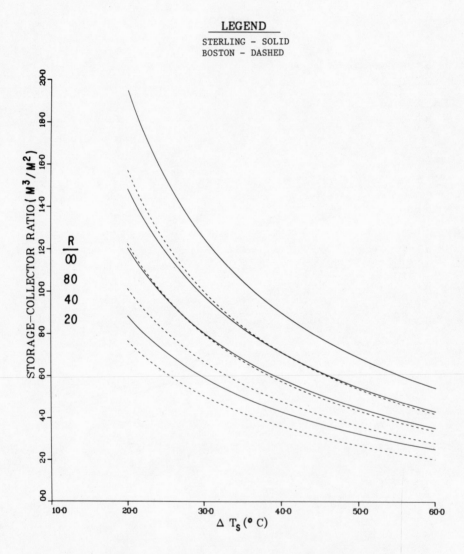

Fig. 3C-D. Ratio of Storage Volume to Collector Area, and
Overall Solar System Efficiency for Sterling and
Boston as a Function of ΔT_S. The four curves,

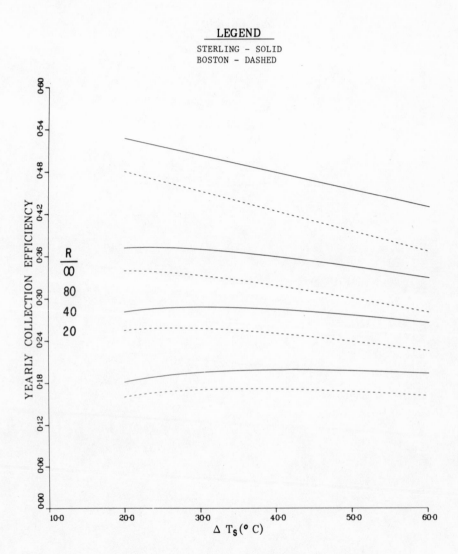

from top to bottom respectively, correspond to
different levels of tank insulation, R = ∞, 80,
40, 20.

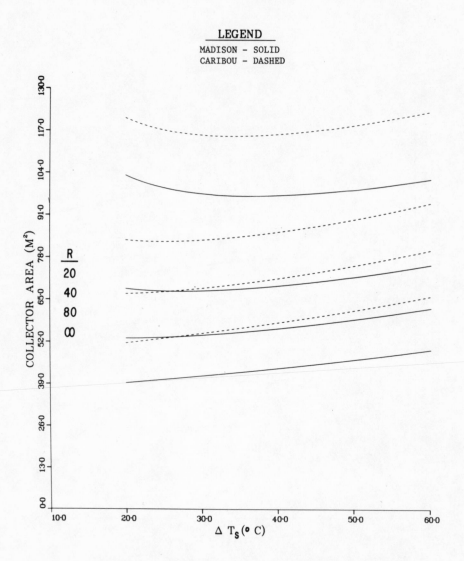

Fig. 4A-B. Collector Area and Storage Volume Requirements
in Madison and Caribou as a Function of Yearly
Variation in Storage Temperature ΔT_S, calculated

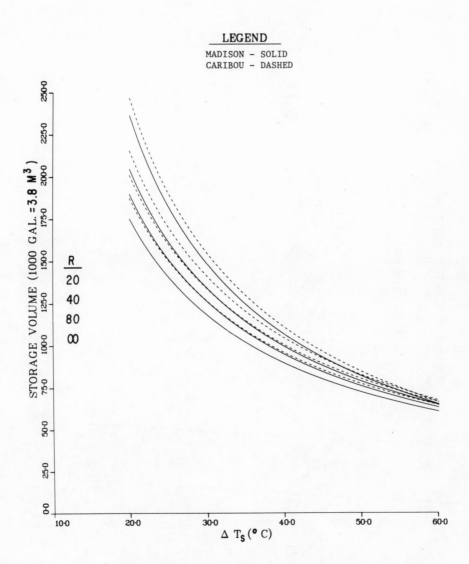

from Eqs. 24-25. The four curves, from top to
bottom respectively, correspond to different
levels of tank insulation, R = 20, 40, 80, ∞.

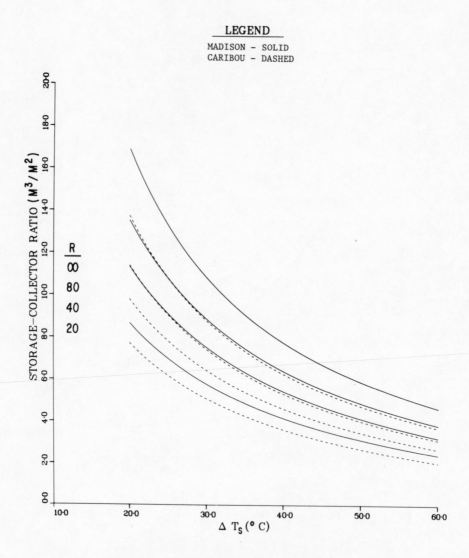

Fig. 4C-D.   Ratio of Storage Volume to Collector Area, and
            Overall Solar System Efficiency for Madison and
            Caribou as a Function of ΔT_S.   The four curves,

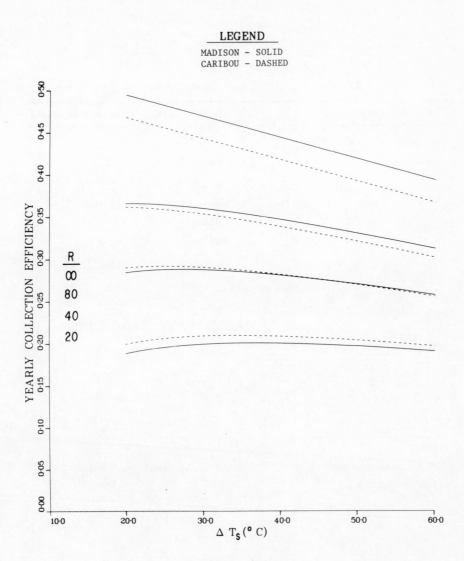

from top to bottom respectively, correspond to
different levels of tank insulation, R = ∞, 80,
40, 20.

Table 3.   System, Load and Cost Parameter Values

Parameters		Metric Units	Parameter Values
SYSTEM:			
Collector	u_c	kJ/hr-°C-m^2	11.
	$\tau\alpha$	dimensionless	.73
	tilt	degrees	latitude +10°
	f	dimensionless	.33
	U_c	kJ/hr-°C-m^2	3.7
Storage	U_t	kJ/hr-°C-m^2	1.-0.(R=20.-∞)
	T_g	°C	13.
	ρc_p(water)	kJ/°C-m^3	4184.
LOAD:	U_ℓ	kJ/hr-°C-m^2	3.
	A_ℓ	m^2	350.
	$(\dot{Q}_\ell)_{max}$	kJ/hr	30-50,000
	$(\Delta\dot{Q}_\ell)_{yr}$	kJ	80-150 × 10^6
COST:			
Capital Charge Rate	i	dimensionless	.10
Fuel Levelization Factor	L	dimensionless	1-2
	c_{conv}	$/kW	50.
	c_F	$/10^6 kJ	3-10
	c_c	$/m^2	0-200

of 2-6. At the higher tank insulation levels, the overall
yearly efficiency of the system is high, reflecting the high-
grade collector parameters used in the calculations. The use
of lower-grade collector parameters would degrade the overall
system efficiency, particularly at the higher values of ΔT_S.

As ΔT_S increases, the effect on area requirements is
twofold: at low values of ΔT_S the effect of increasing ΔT_S
is to decrease A_c because of reduced storage volume, hence,
storage losses; at the higher values of ΔT_S the effect is to
eventually increase A_c because of greater collector inefficiencies. For the range of practical values of ΔT_S (below
60°C) the second effect is seen to remain small for the high
grade collector considered here. As ΔT_S increases the corresponding effect on storage volume requirements is to lower
V_S, roughly as $(\Delta T_S)^{-1}$. In the following section we explore
the economic tradeoff between collector area and storage
volume requirements.

5. Economic Analysis

Using the collector and storage sizing relations
derived in the previous section (Eqs. 24 and 25), we present
in this final section a cost comparison of seasonal systems
versus both diurnal solar systems and conventional heating
devices. As in Section 2, the comparisons are cast in the
form of unit storage break-even costs $\tilde{c}_s$ and $\tilde{c}_s'$, as given by
Eqs. 5 and 7. At storage costs below the break-even value,
the seasonal system has an economic edge over either the diurnal or conventional system.

Adopting the notation in Section 2 the total capital
cost of the seasonal solar system is denoted $c_c A_c + c_s V_s$.
For simplicity the costs of pumps, pipes, controls and heat-
exchangers are assumed allocated to either the collector or
storage unit. Since under reference year conditions the seasonal system is designed to meet the entire load without
additional fuel, annual heating costs are simply given by the
fraction i of total capital costs, with i the capital charge
rate. The annual costs of heating with conventional systems
are the same as specified in Section 2.

A diurnal solar system is conventionally designed to
meet only a fraction of the total annual heating load, with
the remaining fraction met by a full-size backup heating
device. Collector area and storage volume requirements A_c',
V_s' are presented in Table 4 for both 50% and 75% solar dis-
placement fractions (11). For these small diurnal storage
capacities, we set the unit storage cost at the nominal value
of 50¢/gal ($130/m^3). (Because of economies of scale, the

Table 4.   Collector and Storage Sizing Requirements for Diurnal Systems[a,b]

Location	50% Solar Fraction		75% Solar Fraction	
	$A_c'(m^2)$	$V_s'(m^3)^c$	$A_c'(m^2)$	$V_s'(m^3)^c$
Sterling	28.	2.3	58.	4.6
Boston	37.	3.0	76.	6.1
Madison	43.	3.4	93.	7.4
Caribou	46.	3.7	101.	8.1

[a]Collector area sizing obtained from Ref. 11.

[b]Building thermal load, as specified in Table 3, is 1050 kJ/hr-°C (13,300 Btu/Degree Day).

[c]Storage volume to collector area ratio held fixed at 0.08 m^3/m^2 (2 gal/ft^2).

seasonal system can be expected to have a much lower unit storage cost.) Besides the annual ownership costs of the diurnal solar system plus backup device there will in this case be the supplemental fuel costs.

Figure 5 illustrates for Madison the behavior of $\tilde{c}_s$ and $\tilde{c}_s'$ as a function of the system parameter ΔT_s, with the values A_c and V_s calculated from the sizing relations Eqs. 24-25. In this and the following figures we break with metric conventions adopting the more familiar storage cost indice of ¢/gal (¢/gal ≡ $2.64/$m^3$). The cost parameters c_c and Lc_F were set at the values $100/$m^2$ and $6/10^6$ kJ, respectively, and the 50% diurnal system was selected for the comparison. While $\tilde{c}_s$ and $\tilde{c}_s'$ rise directly with ΔT_s for the higher tank R values, they quickly saturate and may actually be negative for the less well-insulated tanks, reflecting the substantial fraction of collector gains consumed by storage losses.

Figures 6A-D give the behavior of $\tilde{c}_s$ and $\tilde{c}_s'$ at each site over a wide range of collector costs c_c and conventional fuel costs Lc_F. To be a cost-effective technology, seasonal storage costs must come in below the minimum of either $\tilde{c}_s$ or $\tilde{c}_s'$.

As the final topic in this section we consider the alternatives available for meeting load under "worst case" winter conditions. Table 5 provides a summary description of the statistical variation in both yearly heating degree days

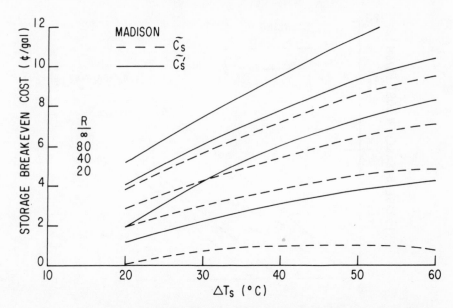

Fig. 5. Storage Break-Even Costs $\tilde{c}_S$ and $\tilde{c}_S'$ for Madison as a Function of ΔT_S for four levels of tank insulation, $R = \infty$, 80, 40, 20. In these calculations collector costs c_C were set at $\$100/m^2$ and levelized fuel costs Lc_F at $\$6/10^6$ kJ.

and yearly insolation levels observed at each site over the last several decades. While the seasonal solar designs specified in the previous sections will meet the entire load during winters milder than the reference year they will fall short during winters more severe than the reference year. The alternatives we consider for making up this load deficit are to oversize the solar system or incorporate a small auxiliary backup system. Although not explicitly considered here, one important factor favoring the need for an auxiliary backup occurs during the initial startup period for the system when it will not be fully charged, requiring some level of backup energy.

The required oversizing of the seasonal solar system is calculated here under the following assumptions: (1) the relative sizing of collector area to storage volume remains constant at the value calculated for the reference year; (2) system efficiency remains constant as the system is oversized; (3) systems are oversized to meet a two-standard deviation in winter conditions or a 1 in 20 year outage probability (95% reliability); and (4) there is no thermal carry-over from year to year, a reasonable approximation

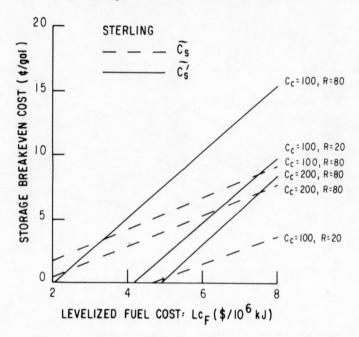

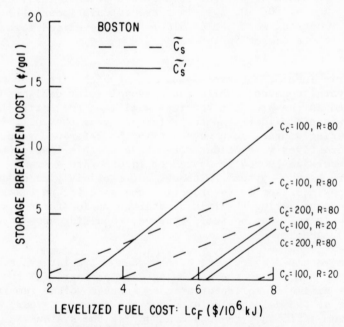

Fig. 6A-B.   Storage Break-Even Costs $\tilde{c}_s$ and $\tilde{c}_s'$ for Sterling
            and Boston as a Function of Conventional Fuel
            Costs Lc_F, for Selected Values of Collector Costs
            c_c and Tank Insulation R Factors.

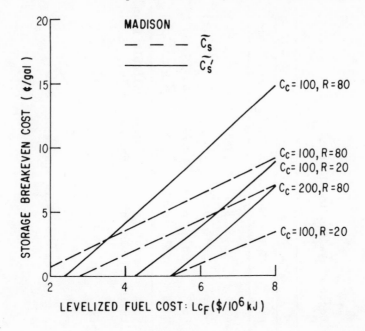

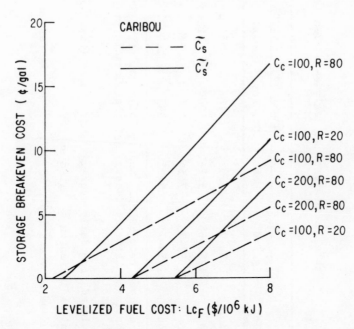

Fig. 6C-D. Storage Break-Even Costs $\tilde{c}_s$ and $\tilde{c}_s'$ for Madison and Caribou as a Function of Conventional Fuel Costs Lc_F, for Selected Values of Collector Costs c_c and Tank Insulation R Factors.

Table 5. Statistical Analyses of Yearly Degree Days and Insolation[a]

Location	Observation Period	Annual Degree Days[b]				TILT	Annual Insolation (10^6 kJ/m^2)[c]			
		Mean	St. Dev.	Max.	Min.		Mean	St. Dev.	Max.	Min.
Sterling	1953–75	4795.	418.	5517.	4084.	45°	5.97	0.23	6.59	5.61
Boston	1953–68	5816.	400.	6312.	4943.	53°	5.60	0.23	5.91	5.23
Madison	1953–75	7454.	404.	8424.	6662.	53°	6.09	0.24	6.43	5.71
Caribou	1953–75[d]	9425.	395.	10000.	8539.	53°	5.61	0.26	5.91	5.10

[a]Data derived from SOLMET, Hourly Solar Radiation-Surface Meteorological Data Base, available from the National Climatic Center, National Oceanic and Atmospheric Administration, Asheville, North Carolina, Dec. 1977.

[b]Calculated from daily average temperatures (base 65°F).

[c]Global radiation on a tilted surface.

[d]Exclusive of years 1961, 63, 67, 68, 73, 74 for which there were gaps in insolation data.

because during late summer/early fall the collectors largely
feed the parasitic losses. For the reference year and more
milder winters we have the following inequality between system output and load requirements

$$0 \leq g \equiv A_c \varepsilon (\Delta H_T)_{yr} - (\Delta Q_\ell)_{yr}, \tag{27}$$

where A_c and ε are the values calculated in the previous
section. Assuming no correlation between yearly insolation
and degree days, the percent increase in collector area ΔA_c
required to maintain Inequality 27 under a two-standard
deviation in winter conditions, that is to maintain the
inequality $g - 2\sigma_g \geq 0$, is

$$\frac{\Delta A_c}{A_c} = \frac{4\sigma^2/x^2 + \left[1 - (1 - 4\sigma^2/x^2)(1 - 4\sigma'^2/x'^2)\right]^{\frac{1}{2}}}{(1 - 4\sigma^2/x^2)} \tag{28}$$

where x and x' refer to $(\Delta H_T)_{yr}$ and $(\Delta Q_\ell)_{yr}$ and σ and σ' are
the standard deviations of $(\Delta H_T)_{yr}$ and $(\Delta Q_\ell)_{yr}$, respectively.
From the data given in Table 5 the percentage increases are
calculated to vary from 15-20% for all four sites.

Assuming the seasonal systems are just cost competitive
with the conventional systems under reference year weather
conditions, a 15-20% increase in capital cost for the seasonal
solar system required to meet worst case winter conditions
appears to be well above the cost of a small auxiliary backup
device plus auxiliary fuel requirements. Because the available storage capacity of the system can be used to smooth out
load patterns, the auxiliary backup can be sized well down
from the design requirements of conventional systems, and
hence be of lower cost.

Nomenclature

Notational conventions adopted in the paper are listed
below, along with metric dimension statements. Generally,
equations in the text are valid in either (consistent) metric
or British units. Unless specified otherwise, all temperature variables refer to daily average values.

A_c	collector area of seasonal solar system, m^2
A_c'	collector area of diurnal solar system, m^2
A_ℓ	exterior area of building, m^2
A_t	storage tank area, m^2
β	"off-season" storage efficiency

c_c collector capital costs per unit area, $/m^2$

c_{conv} unit capital cost of conventional heating system, $/kJ/hr$ or $/kW$

c_F conventional fuel costs, $/10^6$ kJ

c_p specific heat of storage medium, kJ/kg-°C

c_s storage capital cost per unit volume, $/m^3$ or ¢/gal

$\tilde{c}_s$ storage break-even capital cost: seasonal solar versus diurnal solar, $/m^3$ or ¢/gal

$\tilde{c}_s'$ storage break-even capital cost: seasonal solar versus conventional heating, $/m^3$ or ¢/gal

δ $\equiv$ $A_t U_t / A_c U_c$

ε solar system efficiency

f   fraction of day that collectors operate

F_R collector heat removal factor

γ $\equiv$ $A_\ell U_\ell / A_c U_c$

H_T daily average insolation on collector surface, kJ/hr-m^2

i   annual capital charge rate

L   fuel levelization factor

λ_i $\equiv$ $A_i U_i / V_s \rho c_p$

ν fraction of year that "square wave" load persists

ρ storage mass density, kg/m^3

$\dot{q}_c$ daily average rate of collector output per unit area, kJ/hr-m^2

ΔQ_c daily heat gain from collectors, kJ

ΔQ_ℓ daily heat load, kJ

$(\dot{Q}_\ell)_{max}$ peak rate of heat transfer to load, kJ/hr

ΔQ_t daily tank heat loss, kJ

σ standard deviation

t   time, hr

Δt $\equiv$ 24 hours

$(\Delta t)_{yr}$ one year

t_{max} time of maximum storage temperature, hr

t_{min}	time of minimum storage temperature, hr
$\tilde{T}$	temperature difference
T_A	ambient temperature, °C
T_c	effective collection stagnation temperature, °C
T_{cin}	collector fluid inlet temperature, °C
T_g	ground temperature, °C
$\tilde{T}_\ell$	effective load temperature difference, °C
T_r	room temperature, °C
T_s	storage temperature, °C
T_s^{max}	maximum yearly storage temperature, °C
T_s^{min}	minimum yearly storage temperature, °C
$\overline{T}_s$	yearly average value of storage temperature, °C
ΔT_s	yearly variation in storage temperature, °C
$\tilde{T}_t$	effective tank - surroundings temperature difference, °C
$\tau\alpha$	collector transmittance absorptance product
u_c	collector heat conductance per unit area, kJ/hr-°C-m^2
U_ℓ	average building heat conductance per unit area, kJ/hr-°C-m^2
U_t	tank heat loss conductance per unit area, kJ/hr-°C-m^2
V_s	storage volume, m^3
< >	daily average value
< >$_c$	average over collection period

Acknowledgments

We thank Dr. Ben Chin Cha and Mr. Charles Maslowski for their assistance in the numerical computations. We also thank Ms. Mary Lou Bluth for her diligent typing effort.

References and Notes

1. Work supported by the Energy Storage Systems Division, U.S. Department of Energy.

1a. For single building applications, see for example, F.C. Hooper and C.R. Attwater, "Optimization of an Annual Storage Solar Heating System over Its Life Cycle," Proceedings of the International Solar Energy Society Meeting, American Section, Orlando, Florida, Vol. 1, June 1977. Besides single buildings, seasonal storage is under consideration for use in large community-scale applications, both with and without a solar input. See, for example, W. Richard Powell, "CASES Simulation," also in this volume.

2. For a discussion of the relative complexities and costs of augmenting diurnal solar heating systems by electric, natural gas or oil backup devices see, for example, J.G. Asbury, R.F. Giese and R.O. Mueller, "The Interface with Solar: Alternative Auxiliary Supply Systems," Proceedings of the International Solar Energy Congress, New Delhi, India, Jan. 1978. See also J.G. Asbury and R.O. Mueller, "Solar Energy and Electric Utilities: Should They Be Interfaced?", Science 195, 445 Feb. 1977.

3. The capital charge rate i represents the annual cost of ownership. For simplicity we assume that all devices have the same expected life so that i remains constant across technologies. The fuel levelization factor L takes account of real increases, if any, in fuel prices over the life of the device.

4. Our analysis neglects the performance gains achievable with storage stratification as well as by controlled valving alternatives that permit direct transfer of heat between collector and load during the early part of the winter heating season. Although these effects will improve overall system performance, perhaps substantially under some circumstances, they are not expected to invalidate the qualitative sizing and cost conclusions reached here. For an analysis of the benefits of stratification in diurnal solar systems, see Donald W. Connor and Ronald O. Mueller, "The Elementary General Theory of

Stratified Solar Thermal Systems," Argonne National
Laboratory Report ANL/SPG-5, Nov. 1978.

5. Valid in a steady state, this approach neglects transient
   effects caused by heating and cooling of the surrounding
   earth. See, for example, F.C. Hooper and C.R. Attwater,
   "A Design Method for Heat Loss Calculations for Inground
   Heat Storage Tanks," Heat Transfer in Solar Energy Systems, ASME Publication of Proceedings of Winter Annual
   Meeting, Atlanta, Georgia, Dec. 1977.

6. Although not considered here, an analysis of partial
   seasonal solar systems supplemented by an auxiliary
   energy backup could proceed directly from Eq. 14.

7. During the start-up period, extending into the first
   operating season, the system is likely to be under-
   charged requiring a temporary supplemental energy source.
   This initial value problem is not considered explicitly
   in the present analysis.

8. Monthly normals of heating degree days (base 65°F)
   obtained from "Climatography of the United States (by
   State)," Publication No. 81, National Climatic Center,
   Asheville, North Carolina, 1973. Insolation data ob-
   tained from "Application Engineering Manual," Solaron
   Corporation, Commerce City, Colorado, Dec. 1976.

9. In our analysis, collector tilt was fixed at approximately latitude +10° (see Table 1). No attempt was made
   to optimize tilt angle. However, it is clear that because of collector and storage losses the value of a
   solar Btu will increase with its proximity in time with
   the winter heating season, thereby implying tilts that
   favor the winter season.

10. For simplicity, we approximate $<T_A>_c$ in Eq. 9 by its
    daily average value, and assume f is a constant over the
    year.

11. Collector area sizing results were obtained from Pacific
    Regional Solar Heating Handbook, Los Alamos Scientific
    Laboratory, 2nd edition, Nov. 1976, U.S. Superintendent
    of Documents, Washington, D.C. The ratio of storage
    volume to collector area was held fixed at 0.08 m^3/m^2
    (2 gal/ft^2).

COMMENTS

C.J. Swet (U.S. Department of Energy): It was mentioned that the scope of the study was limited to seasonal storage of hot water in constructed tanks for solar heating of a single family dwelling. I agree with the conclusion that seasonal storage is marginally economical within that narrow context, but I suspect that many readers might remember the conclusions without the important caveats on its applicability. Some points will help place the conclusions in perspective.

1. Consideration of the additional use of storage for year round tap water preheating will make the economics look better.

2. Consideration also of the summer cooling function will have a still greater positive impact.

3. Seasonal storage of hot water in constructed tanks is much more economical for larger buildings, partly because the tanks cost less per unit volume but mainly because the reduced surface/volume ratio cuts down on heat loss.

4. Where natural confined aquifers are available for the storage of heated or chilled water, the cost of seasonal storage for large buildings or communities will be at least an order of magnitude lower.

5. Where suitable aquifers are not available, seasonal storage might prove to be an economical alternative.

6. Seasonal storage by means of reversible chemical reactions such as sulfuric acid dilution offers an alternative to aquifers or earth storage, with heat loss essentially nil because of ambient temperature storage.

From this it should be clear that seasonal storage has a very strong potential for economic viability, whether or not the specific approach studied by the authors makes the grade.

Author's Reply:

By comparing the total cost of supply for a seasonal solar system (minus the storage component) against a conventional heating furnace or diurnal solar system, we have derived breakeven cost estimates of seasonal storage for 100% solar space heating of buildings. These estimates

represent <u>upper bounds</u> on unit storage costs consistent
with the constraint of equal total system costs, and are
not to be interpreted as estimates of <u>actual</u> storage costs.
We are, however, encouraged to hear (points 3, 4, 5, and 6
in above comment) of the promising opportunities for long-
term storage currently under development, which may provide
storage at capital costs substantially below our upper bound
estimates.

While DWH preheating will improve the relative economics
of the overall solar syatem as cited in point 1 above, it
will have little or no effect on the relative economics of
seasonal storage since DHW is a year-round load.

The inclusion of the solar cooling function, as cited
in point 2 above, will in principle reduce seasonal storage
requirements and storage losses by increasing the length of
the load period to cover both the winter and summer seasons.
Provided solar cooling is economical, this will have a
positive, although we believe small, impact on the relative
economics of seasonal storage.

<u>Karl W. Boer (University of Delaware)</u>: What is the current
best estimate for a 50,000 gal. storage tank with R=50 ?

<u>Author's Reply</u>:

Unfortunately, we do not have a straightforward answer
to your question. The estimated minimum cost of constructed
water tanks of the size and insulation you specify has been
a subject of considerable disagreement over the past few
years, with estimates ranging from a low of 5-10¢/gal. to a
high of 30-50¢/gal. Part of the explanation for the very
low estimates are innovative construction materials and
techniques, but unfortunately part of it may also be in wish-
full thinking.

10

A Combined Solar Electric and Thermal House System

Karl W. Böer

Abstract

A photovoltaic-thermal cogeneration system for partial solar electrification and comfort conditioning, interfaced with the electric power utility grid, is described. A heat pump used for air conditioning, and, solar assisted, for supplemental heat. A first experimental house, Solar One, built in 1973, and, implementing some of the features, is analyzed during its first three years of operation (1). Conclusions drawn from its operation led to the development of a first residential cogeneration system which was partially completed in 1976/7. One wing of this residence is "heated" by a low temperature system with essentially zero driving force via a copperfoil embedded in the double outside walls of the house and kept near room temperature by pipes fed with heat from an indoor pool, used for heat storage. This permits the use of low temperature roof collectors with maximum thermal and photovoltaic conversion efficiences. First experimental results are reported and analyzed in respect to performance and payback of first cost.

1. Introduction

The conversion of solar energy into useful energy requires a careful systems analysis to maximize the return on the investment. A wide variety of systems are possible and there is no simple means to generalize this analysis.

However, a few facts provide important guidance to solar systems design:

1. Solar energy is supplied at low energy density ($\lesssim 1$ kW/m^2) and almost evenly (mostly within a factor of two) distributed over most of the

populated areas of the globe.

2. The use of diffuse (skylight) radiation adds a
    marked fraction to the direct component in most
    parts of the world.

3. The efficiency to convert sunlight into useful
    energy is higher at lower collector temperatures.

4. Sunlight may be converted simultaneously into
    different forms of energy (e.g. into electricity
    and heat).

It is the goal for opening up major solar utilization,
to design systems which supply the customer reliably with
energy and amortizes through the cost-savings of replaced
energy in acceptable periods of time.

The amortization time is reduced by increased conversion
efficiency, low deployment and distribution (energy transport)
cost, multiple converter utilization, low storage needs, high
life expectancy and low maintenance. Considering all these
factors, the Solar One System was proposed which seemed to
provide a more cost effective conversion for a substantial
fraction of the total energy market. We will summarize the
general analysis of the Solar One System in Section 2.

In order to obtain the first data, a Solar One experimental house was built at the University of Delaware in 1973.
Some of the main results are summarized in Section 3. Based
on these data and a better knowledge on seasonal and regional
insolation, present and projected component and systems costs,
general economics, and trend analysis, it is appropriate to
re-evaluate the Solar One System. Such evaluation is made
in Section 4.

Consequently, one is now able to suggest a variety of
basic system designs derived and further developed from the
general Solar One concept. An example of such design is
partially implemented in a private residence (retrofitted).
This system is described and first results are presented in
Sections 5 and 6 respectively.

2. The Solar One System

The Solar One system consists of (1, 6, 7, 11):

1. A distributed conversion system with rooftop
    collector deployment.

2. Flat plate collectors with solar cells as collector surface and means to extract heat.

3. An interconnect with the electric power utility with means to feed surplus electric energy into the grid.

4. A modest size battery for short term electric storage.

5. A solar assisted heat pump for summer cooling and auxiliary winter heating.

6. A medium size thermal storage bin to bridge approximately two (2) days of inclement weather.

7. Coolness storage to utilize off-peak electric energy for air conditioning.

The reasons to propose these parts for the Solar One system are summarized below.

Rooftop Collector

Rooftop deployment reduces deployment costs as it uses existing structures for such a deployment and permits substantial credits for such items as roof shingles and their installation. A distribution (rooftop collector) system reduces transport costs of energy. These costs could be substantial for electric energy distribution from the generator plant to the consumer and usually are forbiddingly high for transport of low grade thermal energy.

Collector safety, maintenance, building codes, possible tax penalties for the sale of on-site produced surplus energy and esthetically motivated modification were judged as minor deterrents for the distributed compared to a central solar conversion system.

The market fractionation and a probability of a backlash caused by overselling and inappropriate service in early years of market penetration are judged as possibly major impeding factors.

Hybrid Flat Plate Collectors

Flat plate collectors are proposed for reasons of lower cost collection (no tracking) of the direct and diffuse component of the solar radiation. This provides least sensitivity to slight weather deterioration such as high turbidity, haze or thin stratus clouds and takes full advantage of reflection benefits (intentional from adjacent surfaces or unintentional from surrounding structures or clouds).

A combination of photovoltaic and thermal collection provides effective means to "cool" the photovoltaic cells, hence improve substantially their efficiency and increase their life-expectancy.

The use of photovoltaic conversion in conjunction with a heat pump for summer air conditioning eliminates the need for high temperature thermal conversion. This permits also an increased average thermal collection efficiency in regions where the average collector-to-outside-temperature difference during the heating season is lower than the respective difference during the summer for effectively charging a desorption refrigeration system. This is true for most of the continental USA when using a low temperature heating system (see Section 5).

The tandem use of the collector (hybrid collector) permits a major cost reduction. The same collector fulfills the function to generate electric energy as well as heat, and, for a proper system design, both functions can be satisfied close to their optimum conversion efficiency achievable in a single function system. This is caused (a) by the low temperature operation of the photovoltaic cells and (b) by the fact that these cells act as selective absorbers, with most of the light transmitted through the transparent cover being absorbed by the cells, but little being emitted in the long wave length range (band gap emission is negligible).

Interconnect with Utility Grid

Photovoltaic conversion is instantaneous and needs means to match the highly fluctuating supply and demand (10). The most elegant way to achieve this is to actively interconnect the photovoltaic generator with the utility grid and to pump surplus energy into the grid or to use utility-supplied energy when the demand exceeds the solar supply (7). The interconnection provides diversity between a large group of customers.

To properly account for the bipolar flow of energy, forward and reverse metering is proposed.

Electric Energy Storage

A modest amount of electric energy storage is necessary for a variety of reasons:

1. to provide simple means for (close to) maximum power point tracking* of the solar array with varying insolation (nearly fixed voltage),

2. to permit simple dc to ac inversion with storage during off-cycles,

3. to smooth out short supply/demand transients, such as caused by fast moving clouds, airplanes, starting motors, etc.,

4. to provide modest storage for modest* utility peak shaving and load leveling,

5. to have distributed utility storage capacity as short term fall back for black-out emergencies. With proper load management, probably radio controlled from the utility, charging and discharging of these batteries can be coordinated so as to provide some load leveling and systems reliability.

The size of each battery depends on techno-economical development of improved batteries and is presently estimated to carry a few hours load of each customer.

Solar Assisted Heat Pump

There are time periods of marginal weather during which the transport fluid is heated in the collector to temperatures below the storage or room temperature, hence is not useful for direct utilization. The heat pump can properly amplify this thermal energy to useful levels (12).

Long-term projection makes direct use of fossil fuels for heating purposes by the end customer more and more unlikely. Hence, utilization of electric energy for auxiliary heating, however, using a heat pump for efficient overall conversion becomes increasingly attractive.

The use of the heat pump for air conditioning in conjunction with solar cells eliminates the need to purchase another piece of equipment (desorption refrigerator). An estimate of the overall systems efficiency indicates that with solar cells exceeding 7% efficiency, the solar cell/ heat pump air conditioner may exceed the systems efficiency of a solar thermal collector desorption refrigeration system.

The intuitive desire of designers for solar energy

---

*Proper sizing for maximum cost-efficiency is paramount (9).

conversion systems to utilize the amply available heat during
the summer for cooling purposes seems to be misleading, as
the example above shows, when a photovoltaic system of
sufficient efficiency becomes available.

Heat Storage

With a heat storage bin bridging approximately two (2)
days of inclement weather condition in most parts of the
continental USA, an otherwise properly designed system can
supply approximately 75% of the thermal energy demand for
space heating.

Because of the desire to operate the solar collectors
at the lowest possible temperatures, a heat of fusion
storage system is desirable with a solid melting only slightly
below the design temperature for panel operation. Sodium-
thiosulfate pentahydrate is such a salt and is used in Solar
One (1). The melting temperature is 48.9°C. The size of the
storage bin with 250,000 kcal is less than 6 m^3.

Coolness Storage

In regions of high imbalance of energy demands (excessive
daytime peak loads) with expected low night time electric
energy rates (low fuel cost of electric power utilities), it
may be worthwhile to contemplate some operation of the heat
pump during night hours and charging a coolness storage bin.
This stored coolness may then be used to augment the solar
operation of the heat pump during day hours.

Because of the low outside temperatures, the heat pump
will also run more efficiently during night hours (1).

However, a careful return on investment analysis will
decide whether such a coolness storage can be justified. In
Solar One, it was incorporated for experimental and not for
economic reasons.

3.   Solar One Performance*

Most of the experimentation and evaluation of the Solar
One performance was done during the first three years of its
existence.** During most of that time, Solar One's living

---

*Partially reported in (1).
**Since 1976, the author is no longer responsible for the
operation and research done in Solar One. Hence, in this
section's report, the past tense is used when recent modifications altered the original structure, system or intent.

area was unoccupied. Most of the extensive experimentation
took place in the attic and basement. There was no occupancy
simulation. However, considerable traffic of visitors was
encountered (average of 200 persons per week).

The experimentation was mainly directed to obtain performance data of each component of the system, to improve
the flat plate collectors and to optimize the system. During
such experimentation, the system usually yields minimum performance information, since an average over operating and non-
operating periods is taken. (Because of the heat capacity of
the house, it is impossible to separate properly rapidly
following periods of operation and shut-down). Because of
the various experiments performed, a great variety of operational modes were tried out with a variety of components
employed. In the following description, only typical or
nearly optimized modes and components are described even
though they may have been employed only for a short fraction
of time.

Solar One Construction Data

The first floor plan and cross section is given in
Fig. 1. A front view of the house is shown in Fig. 2, with
24 roof collectors (1.20 x 2.40 m^2 (4' x 8') each) on the 45°
angled roof facing 4.5° west of south. Six south wall
collectors have an active area of (1.20 x 1.80 m^2 (4' x 6')
each).

All house walls have sprayed-in-place 2.5 cm (1") K 13
Type T (National Cellulose Corporation) insulation plus two
layers of 9 cm (3 1/2") fiberglass blankets, one side aluminum
foil coated, installed back-to-back between 5 x 10 cm^2
(2" x 4") studs. The entire wall area is then covered with
a four mil polyethylene film and finished with sheetrock,
tapejoint, spackeled and painted. The outside of the house
has Redwood bevel siding on top of a 2.5 cm (1") Armstrong
sheathing.

The heat loss rating of the house is 4900 kcal/DD(°C)
(3 kWh/DD(°F)) for 21°C living space temperature as determined from the estimated heat transfer through all walls and
by measurements during a sequence of five cloudy winter days.

Solar Heating System

The solar collectors are placed from the inside against
a 1.25 cm (1/2") Abcite-AR coated acrylic skylight between
5 x 25 cm^2 (2" x 10") roof rafters spaced 1.2 m (4') on
centers. The skylight is sealed with butyl rubber and

Thiokol resilient cement. The collectors are contained in a plywood box and have a total active surface of 56 m^2 (620'^{2}) (roof) and 12 m^2 (130'^{2}) (southwall). Each box is covered with a 3 mm (1/8") PPG white glass or plexiglass. The collector plate is an aluminum metal sheet coated with flat black paint (for finning of this plate see Section 3). The collector box contains 5 cm (2") polyurethane foam insulation on the back and side wall and provides a 4.4 x 104 cm^2 (1 3/4" x 41 1/2") air duct beneath the absorber space (see Fig. 3).

The air flow through each collector pair can be regulated separately in order to adjust the system for equal collector outlet temperatures.

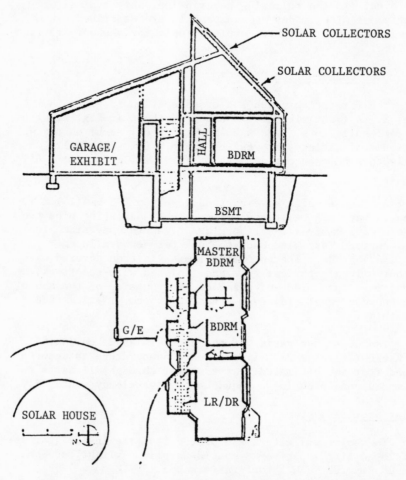

Fig. 1 Solar One - Floor Plan and Cross Section

Fig. 2   Solar One house at the University of Delaware

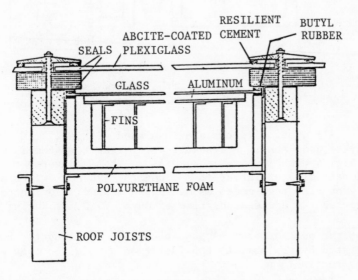

Fig. 3   Cross section to roof collector in Solar One

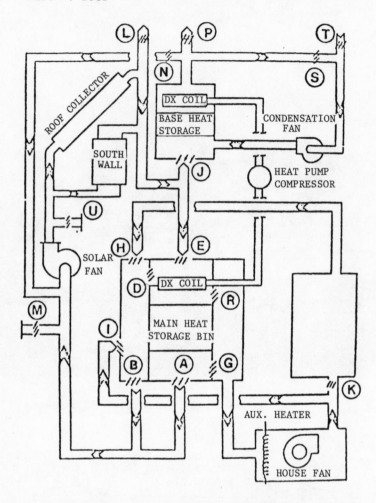

Fig. 4   Flow diagram with louvers A-U and fans

All other airducts are insulated with a 2.5 cm (1")
polyurethane sheet. A schematic of the duct system is given
in Fig. 4 with louvers labeled A-U. These louvers are
pneumatically operated* from a control system, switching
between different modes of operation depending on certain
temperatures and seasons/time of day. E.g., if during the

_____

*Not the most economical way to operate the louvers, but
dictated by reliability consideration during the time of
construction.

heating season, the temperature of the air coming from the
collector is 10°C larger than the temperature of the heat
storage salt in the main bin, louver E and A in the collector
circuit are open; I, D, R, B and G are closed. When the house
calls for heat, H, B, R, and K are also open; etc. There
are more louvers than necessary for house operation. These
louvers are used for certain tests. E.g., the heat storage
bin may be charged by resistance heating with louvers I, D
and G open and H, E, R, A, B and K closed.

The main heat storage bin as used during these experiments is shown in Fig. 5. Air could be either directed
through the space between the center tubes or through the
space between the outer panels, avoiding a short path through
the inactive part (i.e., the 48.9°C part during summer and
the 12.8°C part during winter operation), however, with some
benefit from the specific heat of the inactive material.
Dimensions and heat capacity of the main storage bin are
given in Table 1.

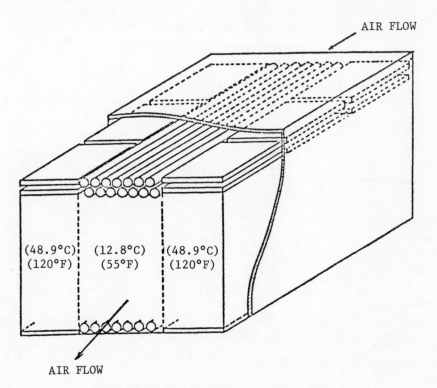

Fig. 5 Thermal energy storage bin

Table. 1. Dimensions and Thermal Data of Thermal Energy Storage Bin

MAIN THERMAL STORAGE BIN		
Storage Material	$Na_2SO_4 \cdot 10H_2O/NaCl/Nh_4Cl$	$Na_2S_2O_3 \cdot 5H_2O$
Melting Point	12.8°C	48.9°C
Specific Heat solid	0.35 kcal/kg°C (Btu/lb°F)	0.35 kcal/kg°C (Btu/lb°F)
Specific Heat liquid	0.65 kcal/kg°C (Btu/lb°F)	0.60 kcal/kg°C (Btu/lb°F)
Heat of fusion	43.4 kcal/kg (78 Btu/lb)	50 kcal/kg (90 Btu/lb)
ABS Containers:		
Unit/dimensions	3.13 cm (1.25") O.D. 180 cm (6') long	$2.5 \times 53 \times 53$ cm^3 (1"x21"x21") pans
Wall thickness	0.75 mm (0.030")	1.25 mm (0.050")
Surface area	0.18 m^2 (1.96 ft^2)	0.59 m^2 (6.6 ft^2)
Empty weight	0.16 kg (0.36 lb)	1.01 kg (2.25 lb)
Contents weight	1.81 kg (4.02 lb)	10.7 kg (23.8 lb)
Number of containers	620	294
Total container surface	110 m^2 (1200 ft^2)	170 m^2 (185 ft^2)
Specific heat of ABS	0.4 kcal/kg°C (Btu/lb°F)	0.4 kcal/kg°C (Btu/lb°F)
Storage Bin:	5.83 m^3 (2.6 ft^3)	
(% air passage)	56%	32%
Total Heat Stored		
As heat of fusion	49,120 kcal (195,000 Btu)	158,700 kcal (630,000 Btu)
As specific heat		
Salt 21°C to 49°C	15,400 kcal (61,000 Btu)	30,700 kcal (122,000 Btu)
Salt 49°C to 52°C	2,270 kcal (9,000 Btu)	6,300 kcal (25,000 Btu)
Containers 21°C to 10°C	1,800 kcal (7,200 Btu)	3,700 kcal (14,800 Btu)
TOTAL heat in Bin 21°C to 52°C (20% of 55° (12.8°C) bin is added)		203,300 kcal (807,000 Btu)
Total Coolness Stored		
As specific heat		
Salt 24°C to 13°C	8,200 kcal (32,500 Btu)	12,300 kcal (49,000 Btu)
Salt 13°C to 10°C	1,100 kcal (4,400 Btu)	3,100 kcal (12,300 Btu)
Containers 24°C to 10°C	800 kcal (3,200 Btu)	1,700 kcal (6,600 Btu)
TOTAL coolness·in Bin 24°C to 10°C (10% of 48.9°C bin is added)	61,000 kcal (242,000 Btu)	

The air was circulated through the collector fan of 1.5 hp max. rating (1.5 kW motor), usually at a rate of about 100 m^3/min (3600 cfm) at 3.8 cm (1.5") H$_2$O pressure drop across the fan. The collector fan motor was connected to a proportional control circuit which reduces the flow rate during periods of lower insolation to provide a solar panel outlet air temperature of 10° in excess of the heat storage salt temperature at the bin outlet. The average hourly energy consumption of the fan during operation was 0.8 kWh during times of maximum collection.

The air through the house was circulated at a rate of 32 m^3/min (1200 cfm) with 7.5 mm (0.3") H$_2$O pressure drop, powered by a fan with a one-half hp motor (0.52 kW).*

The auxiliary heating system consists of a 3 ton (2.9 kW*) heat pump (York Airconditioning Corporation) and 3/4 hp condensor fan (0.62 KW*) which delivers 65 m^3/min (2400 cfm). The heat pump/condenser fan combination operated when the main storage bin temperature dropped below 25°C. An 8 kW resistance heater was used when the temperature at the condensor fan dropped below 4° C.

Solar Electric System

Three solar roof panels are filled with CdS/Cu$_2$S solar cells manufactured between 1968 and 1970 by the Clevite-Gould Corporation. Always 104 of these cells, selected for the best short circuit current match, are connected in a series in a subpanel and cemented onto a finned galvanized iron sheet. This sheet is then covered with a front lucite sheet and hermetically sealed at the edges leaving a 1.3 cm (1/2") airspace above the cells. Three of these subpanels are assembled to a 1.2 x 2.4 m^2 (4' x 8') (nominal) panel. The space above the cells in each subpanel was continuously flushed with dry high purity nitrogen. Several of these subpanels were electrically loaded near the maximum power point with an external resistor and their output was regularly monitored. Other subpanels were connected directly to a lead acid battery. These batteries also have the effect of stabilizing the output voltage at a value close to the maximum power point during variable insolation.

The dc system could be directly connected to the 110/220 V dc load system of the house by a main transfer switch.

---

*The electric power consumption is given in normal full load condition.

A dc to ac inverter (solid state, of 3.0 kW power rating or motor/generator of 3.0 kW max.) could be used to provide energy to typical ac consumers (refrigerators, motors, TV, etc.).

For charging-discharging experiments, a power supply slaved to a solar cell panel array was used, simulating total roof coverage with CdS/Cu_2S solar cells.

The energy consumption of the different key elements of the solar system was measured and recorded daily in 8 separate meters.

Airconditioning

For airconditioning purposes, the heat pump is used. It could be powered by solar electric energy after inversion. The heat pump in Solar One was operated primarily during night hours with off-peak electric energy from the power utility and the thermal energy ("coolness") was stored in the main bin at 12.8°C.

Solar Heating Systems Performance

A proper matching of collector with heat storage and user (living space), including optimization of transfer fluid transport, is essential for efficient collection and conversion. In the Solar One system, this means that special attention had to be given to

a)   minimizing the temperature differences between heat collecting surfaces (solar collector or storage bin) and air, so as to permit a low collector surface temperature while still melting the storage salt at 49°C;

b)   minimize collector and bin cost and air resistance, hence minimize finning of these surfaces;

c)   select proper air volume and airflow patterns to transport the collected heat with minimum losses and with minimum control elements (cost) to obtain highest efficiencies.

Since little experience and data were available, the route of experimentation was chosen with a first, rather simple system installed and stepwise improved.

With the specific heat of air of 285 cal/°C m^3 and ca. 39,000 kcal per hr generated at the roof and south wall (65% collection efficiency), a flow rate of 65 m^3/min is needed if a temperature rise within the collector of 10°C is permitted.

Since additional losses are unavoidable and an additional substantial ΔT is necessary as driving force for collector surface-to-air and air-to-bin surface, such a high temperature swing is barely acceptable. The design value was consequently reduced by increasing the airflow to 100 cm^3/min, yielding a temperature swing of ~ 6.5°, however, at the expense of pumping power, P:

$$P(hp) = F(m^3/min) \ \Delta p(cm \ H_2O)/443 \qquad (1)$$

with F the airflow rate and Δp the pressure difference across the pump (friction loss in the air handling system). With $\Delta p \simeq 4.5$ cm H_2O, a pump of 1 horsepower is required. The higher rating of the actual pump was used to provide experimental flexibility.

The parallel roof collector feeding through tapered feeder ducts is relatively easy to balance and provides means to minimize airflow resistance in the collector bank.

Plenum feeding with airflow balance achieved by restraint design within the heat storage bin was used initially for economic reasons.

Solar Collectors. Little data were known about finned collector design. With about 560 kcal/m^{2}h, estimated heat to be transferred from the collector (65% efficiency), it is obvious that an unfinned smooth surface would necessitate a $\Delta T_c \simeq 55°C$ as driving force to heat the air, hence requiring an unacceptable upper collector plate temperature more than 60°C above the melting temperature of the salt (this would render the thermal and photovoltaic conversion efficiency too low).

Permitting one-tenth of this value, $\Delta T_c \simeq 6°C$, one arrives at an economically unacceptable finning surface of 10 times the collector surface, hence indicating the need for rough surfaces.

Experimentation was necessary to obtain information about such surfaces and to achieve a compromise between good heat transfer and acceptable airflow resistance. In the course of these investigations, in which 16 different types of collectors have been tested on the roof of Solar One, it was observed that the heat transfer in such asymmetric ducts* can be described (Nu and Re are the Nusselt and Reynolds numbers

---

*Ducts below the collector surface (see Fig. 3) are bound by the polyurethane foam at the other three sides.

respectively) as:

$$Nu = A_1 \cdot Re^{0.8} \qquad (2)$$

with a substantially larger A_1 than Carter's factor (A = 0.0156) (2). The ratio A_1/A may be described for rough ducts with fully developed turbulent flow as:

$$A_1/A = \frac{Nu_{rough}}{Nu_{smooth}} = \alpha \sqrt{f/f_o} \qquad (3)$$

with f and f_o the coefficient of friction for rough and smooth ducts respectively. According to Kolar (3), $\alpha = 1$ for smooth and rough tubes.

The experimental results with several finned asymmetrical ducts (collectors) indicate that α is approximately equal to the area ratio of the finned to the planar collector surface: S_f/S_o (4). Hence, the heat transfer of such finned collectors can be described as:

$$Nu = 0.0156 \cdot S_f/S_o \cdot \sqrt{f/f_o} \cdot Re^{0.8} \qquad (4)$$

It was also observed that interrupted and staggered fins are sufficient at the given airflow rates to fully develop turbulent flow within less than 20% of the collector length for fins of less than 40 cm length.

From Eq. (4), one may conclude that smooth finned collectors are best since the heat transfer increases proportional to the fin area, and only as the square root to the increased friction, while the friction itself enters linearly in the increased need for pumping power. However, as long as the friction of the total collector field is kept substantially smaller than that of the rest of the airduct system, fin staggering (to obtain effective duct roughness) is permissible without marked penalty.

An acceptable compromise seems to be a design with staggered fins of 8 cm length and 4 cm height, placed 3 cm apart which yields a driving force of $\Delta T_c \simeq 9°C$ and a pressure drop of only 2 mm H_2O at 100 m^3/min airflow across the collector banks. Fig. 6 shows a typical temperature profile along the collector as measured for the plate and the air at a flow rate of 80 m^3/min (2900 cfm) with $S_f/S_o = 3.7$ and $f/f_o \simeq 2.8$.

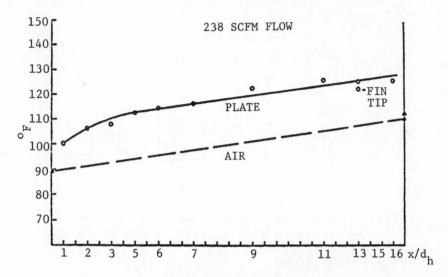

Fig. 6   Temperature profile along the airflow axis
of a finned collector as described in the
text above (d_R = hydraulic diameter).   (4)

Finning profiles and temperature distributions of
examples for too much (a) or too little (b) finnings are
given in Figs. 7 and 8. In the first case, the air friction
is too high ($\Delta p \sim 3$ cm H_2O), in the second case, the driving
force is too high ($\Delta T_c \simeq 17°$).

Summer Operation of Collectors. As indicated in
Section 2, it is of advantage to use a photovoltaic collector
in connection with a heat pump to provide cooling during the
summer months. This system, however, works only satisfactorily when the solar cells are sufficiently cooled, since their
efficiency decreases about 0.3% per degree; hence, the overall
conversion efficiency for the cooling operation can be
improved by ~ 10% by merely reducing the collector temperature
from 90°C to 60°C.

By opening louvers U and L (Fig. 4), chimney action is
employed to reduce the collector plate temperature during
summer operation. Only when the temperature exceeds 60°C,
the solar fan was employed to keep the temperature below 60°C.
For most of the day, the natural convection was sufficient to
maintain the design temperature (5).

The operation at temperatures substantially below 60°C,
however, would require major use of the solar fan. If a

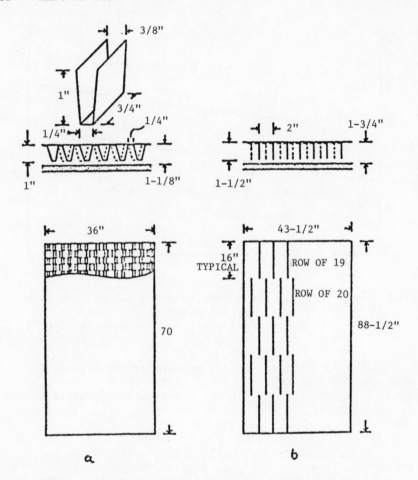

Fig. 7   Cross-section and top view of experimental
collectors with too much (a) and too
little (b) finning.

reduction of the collector temperature by ΔT could be
achieved with a duty factor D of the solar fan, such operation
would be cost efficient only for a minimum systems efficiency,

$$\eta_{syst} = 100 \ P_e D/A_c \Delta T; \tag{6}$$

with P_e = 1.5 kW, the electric power of the solar fan, and
A_c = 56 m^2, the area of the roof collector, the required

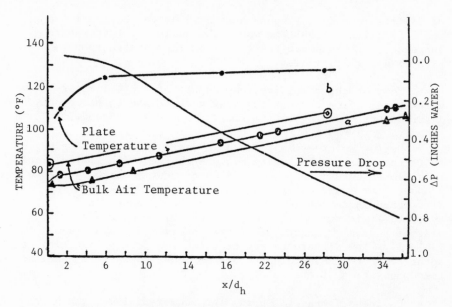

Fig. 8   Temperature distribution for collector
a and *b* of Fig. 7. Pressure drop of
collector *a*.

η_{syst} is 13% for D = 100%, reducing the collector temperature
to 40°C. This indicates that with the existing systems
efficiency of ~5% (using a COP of 2 for the heat pump), it
does not pay to reduce the summer operating temperature of
the collector markedly below 60°C through substantial use of
the solar fan.

Thermal Storage Performance. Because of heat leaks in
the system (mostly through incompletely closing dampers) and
partially faulty salt encapsulation as well as insufficient
instrumentation, the performance data obtained are inconclusive.

Heat Storage. With a heat transfer surface of 170 m^2
and a maximum heat collection at 68 m^2 (roof and southwall)
collector surface of 30,000 kcal/h, one estimates a necessary
driving force of $\Delta T_B \simeq 20°C$. This is substantially larger
than the driving force for the collector and indicates that
the thickness of the pans should be reduced to approximately
1 cm. However, an unacceptable increase in container cost by
more than doubling the number of thinner containers was the
reason for the chosen compromise.

Such compromise at the storage side of the system was
more acceptable since most of the time, part of the heat could

be diverted for direct house heating,* hence only during a
small fraction of the time would the collector temperature
increase slightly above the winter maximum design temperature:
$49.8° + \Delta T_c + \Delta T_B + \Delta T_{losses} \simeq 70°C$.

Most of the time, the temperature of the collectors
were kept** below the maximum design temperature, hence only
insufficient charging of the storage bin was achieved.

During an experimental run in early 1975 employing a
resistance heat, the storage bin was charged with 275,000 kcal
to a temperature of 52°C and was consequently discharged in a
27 hour time span, delivering more than 185,000 kcal. The
difference indicates the magnitude of heat losses in the
system. A minimum of 125,000 kcal of this heat was stored as
heat of fusion or about 80% of the theoretical value (see
Table 1). The difference may be accounted for by insufficient
accuracy of the measurement and some leaky containers changing
the composition of the salt.

Coolness Storage. During the summer of 1974, the heat
pump was run mostly during the night hours. The coolness
storage bin was cycled daily according to the house demand
for air conditioning. Although some reduction in storage
performance was observed during early cycles, essentially no
marked changes were observed later; the amount of coolness
stored was sufficient to keep the house below 24°C during
days when the coolness demand was not excessive.

With heavy traffic of visitors or other excessive
coolness demand, additional operation of the heat pump during
the day hours was required at times.

The total amount of stored coolness after filling the
bin to an outlet temperature of 7°C is 40,000 kcal (150,000
Btu), with approximately 25,000 kcal (100,000 Btu) heat of
fusion contribution. Thermal losses through the bin wall
(4,800 kcal) and through dampers and ducts (4,300 kcal)
accounted probably for most of the losses.

Extended flat ranges in the temperature-time profile
were not observed during heating (1). Some flat portions
during cooldown indicate freezing near 10°C.

_____

*With 50% of the solar heat used directly for house heating,
ΔT_B decreases to 10°C.

**Substantial difficulties to adjust southwall and roof col-
lector for dynamic matching, heat leaks at the southwall collector and use of ordinary black paint prevented efficient
collection at collector temperatures above 60°C.

Heating of Living Space. Direct heating of the living
space from solar collectors or from auxiliary mechanical equipment presents no problem. Extracting heat from the storage
bin requires another ΔT_{bh} as driving force, consequently
delivering heat to the room at a temperature below the melting
point of the salt.

However, since the heat demand of the house at -10°C is
of the order of 150,000 kcal/day, or ~6,500 kcal/h, the
average ΔT_{bh} is only ~4°C, hence the air is delivered at a
temperature of ~45°C from the charged bin. This requires air
circulation at a rate of ~20 m^3/min. The house fan delivers
32 m^3/min., hence is sufficient to extract heat even from a
partially discharged bin.

Electrical Systems Performance

The solar cells used for the panels had a conversion
efficiency at AM1, 25°C of 4.2 to 4.8% before they were
assembled to the collector arrays. After assembly, covering
with two transparent layers (white glass and Lucite AR) and
at the operating temperature (35°C < T_{cell} < 70°C), the average
systems efficiency oscillates between 2 and 3% (see Fig. 9),
depending on the season (H_2O-vapor, solar elevation), temperature, dust cover, and other yet unknown factors (possibly
annealing of some photochemical reaction in CdS).

However, it is important to recognize that within the
experimental error no degradation of the CdS/Cu_2S solar cells
is observed during the rooftop deployment and while the
collectors were flushed with high purity nitrogen (for almost
three years).

The total amount of electrical energy harvested during
typical spring and summer days is 11.9 kW (5-1-1974, 25%
cloud cover) and 14.1 kWh (7-16-1974, 15% cloud cover) with
simulation of coverage of all 24 roof panels with subpanels
identical to the investigated one. Other typical data vary
between 2 kWh (overcast) to 14.6 kWh (essentially clear) with
most of the days (70%) in excess of 6.5 kWh and 50% of the
days in excess of 11 kWh/day.

During the early testing period, 4 subpanels were
connected in series (total of 416 cells in series) and this
array, protected by a diode, connected directly to a series
of 10 car batteries (120 V nominal). The current drawn to
charge the batteries was monitored and through a control
circuit and power supply, an additional amount of charge was

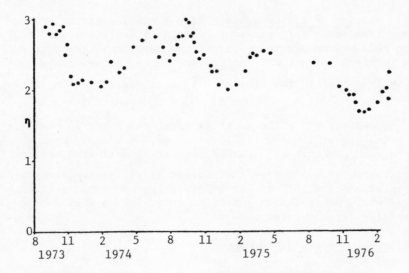

Fig. 9   Efficiency of CdS-panel No. 2-1 on Solar
One during "clear" days determined from
maximum power point to measured total solar
flux ratio (45° slope) above the roof. No
corrections applied for dust, temperature,
reflected light, etc.*

supplied to the batteries determined by the ratio of total
roof collector surface to actual solar cell array area.
Power point tracking by keeping the voltage nearly constant
was satisfactory (within 15% of maximum power point), however
during a sequence of 20 charge-discharge cycles, some uneven
charging of the batteries was detected. Charge/discharge
efficiencies of 85% were estimated.

For operating Solar One's ac appliances, a simple solid
state 3 kW inverter was developed (5) which operated satisfactorily.

4. Solar One Evaluation

Conclusion for Solar One Experiments

The experiments performed in Solar One during the first
3 years have shown that a combined conversion of solar energy

*The jump in efficiency in summer 1975 occurred after the panel
was left to open air during several weeks of repair and removal
of some cells which have developed a short of the solar cell
electrode tab to the finned metal base of the collector.

into heat and electrical energy is technically sound, using a system as given in Solar One, and, when optimized, promises economical attractiveness.

However, most of the components need to be changed in size and a number of structural modifications need to be employed to convert the total system from an experimental structure to a more cost efficient prototype house.

The major results can be summarized as follows:

Air System: The system works satisfactorily, however air leaks at ducts, collectors and bin are costly in loss of energy and not always easily detected. Louvers should be electrically activated, minimized in number, and, where losses are possible, should be insulated and must close tightly.

For reasonable cost of collector and bin, a minimum design temperature of the collector top surface 25°C above the melting point of the salt is required. Proper collector finning is essential to avoid higher collector temperatures.

With properly designed duct geometry, a total pressure drop in the entire system of ~2 cm H_2O may be achievable, requiring a 1/2 HP blower.

A variable speed motor (multiple winding) could be of advantage to maintain the design temperature of the collector and extend the harvesting time.

Interconnection of collectors with different orientation are difficult to dynamically balance and should be avoided.

The house airhandling system should be matched to the auxiliary equipment (heat pump).

Attention should be given to the design of the operating modes and switching logic of the total house system, to avoid wasteful operation.

Heat of Fusion Storage: Such system has a major advantage by using less space and operating at a well defined storage temperature most of the time.

The disadvantage is probably higher cost, loss of gradual charging with marginal heat in stratified bins and possibly limited bin life.

Heat of fusion storage needs substantial further development in respect to material, container and bin structure.

Combined Solar Cell/Thermal Collector: Temperature
stabilization at the design temperature (70°C winter and 60°C
summer) can be achieved. Electrical harvesting with present
generation cells (2-3% systems efficiency) is marginal. With
7% cells and 5.5% systems efficiency approximately 3 kW
electric power with 15-30 kWh (winter-summer) per day
harvesting on the 56 m^2 collector may provide more than 50%
of the demand.

No marked degradation of the CdS/Cu$_2$S solar cells array
over the three years has been observed in the given deployment
mode.

Heat pump: The use of a heat pump as auxiliary equipment for winter heating and as main source for summer cooling
seems to be economically justified.

Proper sizing is essential including operating range,
coil and heat exchanger design to avoid inefficient operation
and summer coil icing.

Proper design of systems operating modes is essential
for economic operation: During the summer, this includes day
(direct) and night (storage and utility) cooling and during
the winter season, this includes marginal heat amplification
direct and storage (during night hours) heating, and operation
in conjunction with a booster heater.

Electrical System: Minor partial electric storage in
a directly connected battery (no maximum power point tracking)
with minor protection seems to work satisfactorily when fed
with low voltage (12 V rather than 120 V).

Load switching (dc and inverted ac to utility ac) is
acceptable. However, active inversion with surplus feedback
into the utility grid may be preferable (single wire system).

Solar One House Critique

The Solar One principle, as developed earlier (6, 7)
has been supported by the results of the experimentation in
the Solar One house. However, even though the Solar One
house as an experimental structure was intentionally built to
provide a wide range for parameter and even component variation had some limitation in its overall system. Most severe
was the choice of an air system and of a heat of fusion
storage bin.

The most obvious result of this choice is still
relatively high collector design temperature (60 rsp. 70°C).

This reduces the overall efficiency of the collector substantially. It is obvious that the overall efficiency of the electrical sub-system can be increased by 10% if it is possible to reduce the design temperature by 30°C. Moreover, the cost-efficiency of the thermal collector sub-system may be increased substantially if such temperature reduction is possible.

The main question is now how to store and utilize such low grade heat economically. If this problem can be solved, a further developed Solar One system may evolve with higher economical attractiveness.

5.   Retrofitted Residence

The problem to develop a system with minimal collector design temperature, however utilizing the low grade heat for living space heating was approached by designing a retrofitted Solar One system at a residence which had already an indoor swimming pool. A large volume of water is needed for heat storage and a relatively small temperature swing, as long as a fully developed heat of fusion storage system is not available. An indoor pool provides such means. It is also available during the entire year and, with a content of 100,000 l (actual size), presents a heat capacity of 700,000 kcal within the comfort range of 22-29°C, sufficient for more than 2 days heat storage for this residence (29,000 kcal/DD (°C) demand for total residence).

The next problem, to heat this residence with the heat from this swimming pool, necessitates substantial heat transfer surfaces. In order to utilize minimal pool temperatures, it was decided to extend such heat transfer surfaces over the entire inside of all outfacing walls and the ceiling. With a surface area* of approximately 510 m^2 and a demand* of 9,700 kcal/DD (°C), one estimates $\simeq$ 2°C as necessary driving force to maintain 20°C room temperature at -10°C outside temperature.

This indicates technical feasibility. It was consequently decided to build such a system in several steps and to proceed with each following step as encouraging experimental results from the previous step are obtained.

First Step: Insulation

A ranch house is used with 265 m^2 base area and an attached 115 m^2 poolhouse. The basement of the main house of

---

*House except poolhouse.

Fig. 10   Aerial view of the hilltop house near
          Kennett Square with solar collector (left)
          and solarized NW part (upper half) of the
          house.

250 m^2 is fully developed. In the attic, two offices and a
small library of 54 m^2 were recently added. It is located on
a hilltop near Kennett Square, Pennsylvania (Fig. 10).

The house was built with 5 x 10 cm^2 (2 x 4") framing
covered with 1.2 cm (1/2") Celutex sheating, 1.2 cm (1/2")
sheet rock and no insulation in the airspace. Its outside is
brick veneer of 2.7 m (9' height). The heat transfer value
of the wall was 0.9 kcal/m^{2}h°C (0.18 Btu/ft^{2}h°F).

The basement was built with 30 cm (1') concrete hollow
blocks and extends 0.6 m (2') above ground (room height in
basement is 2.4 m (8').

The house was heated through hot water ceiling radiation-
heat through pipes imbedded in 2 cm plaster below sheet
rock. The ceiling was insulated in the attic with 9 cm
(3 1/2") fiberglass pads. The house (main floor) contains
80 m^2 thermopane windows, of which 32 m^2 are positioned in
the poolhouse.

The following insulation measures were taken:

a)   18 cm (6") blown-in loose cellulose in the attic was added.

b)   All airspaces between the frame were filled with fiber-glass pads or blown-in loose cellulose.

c)   The basement ceiling was insulated with 9 cm (3 1/2") fiberglass pads where basement rooms are usually not heated (garage, utility).

d)   Direct feeder airducts from the outside are provided for the oil furnaces.

e)   Above ground basement outside walls were insulated with 9 cm (3 1/2") fiberglass pads or 2 cm (4/5") polystyrene.

f)   A curtain wall was installed to reduce losses through a major windowwall of the house (14.5 m^2).

Pre-installation heat loss for an average 2780 DD°C (5000 DD°F) winter*, installation-cost, estimated energy saving/year and payback time for an initial cost of 11¢/1 heating oil (65% burner efficiency) hence $18.80/Mkcal ($4.75/million Btu) are given in Table 2. With a total installation cost of $3,250 and the annual saving of ~ $785, a very reason-able payback time of ~ 4 years is estimated.

A major additional heat loss of the residence, however, was attributed to the evaporation of water from the indoor swimming pool. From July to November, the water level was monitored daily, and a loss of 2.5 ± 0.3 mm/day was determined, yielding 79 kcal/day heat of evaporation loss. A polyethylene foil with motor driven retraction mechanism was devised and reduced evaporation by approximately 95%. As a result, not only a substantial fraction of the heating but also major ventilation cost to reduce the humidity could be eliminated. The payback time of this subsystem is 4 months (Table 2).

Second Step: Solar Pool Heating

The poolhouse has an SSE (175°) facing roof of 30° inclination and 65 m^2 surface. This entire surface was converted into a collector as a first step towards solariza-tion of the residence. The collector was custom made. Figure 11 shows a cross section through the collector. It was built on top of the existing roof shingles using commer-cial greenhouse strutts to hold the single plateglass cover. A 0.5 mm copperfoil with soldered 1 cm (3/8") copper pipe

---

*Microclimate at hilltop location.

Table 2. Payback of Certain Energy Conservation Items

Part of House	Area (m²)	Original/Reduced Heat Transfer Rate (kcal/m²h°C)		Original/Reduced Heat Loss (kcal/year)		Cost Of Installation ($) (C)	Saving Of Energy (E) ($/year)	E/C %	Payback time (years)
A. Attic	380	0.34	0.14	$8.61 \cdot 10^6$	$3.55 \cdot 10^6$	1,044.–	95 (oil) 90 (el)	18	6
B. House Walls (ex.poolhouse)	208	0.90	0.28	$12.5 \cdot 10^6$	$3.88 \cdot 10^6$	675.–	162 (oil) 150 (el)	46	2
C. Basement Ceiling	126	1.63	0.29	$4.9 \cdot 10^6$	$0.88 \cdot 10^6$	352.–	76 (oil) 20 (el)	27	4
D. Air Feeder to oil burner						45.–			
E. Basement Walls	46	1.78	0.35	$4.9 \cdot 10^6$	$0.97 \cdot 10^6$	120.–	74 (oil)	62	1.5
F. Curtain Wall	14.5	2.68	1.0	$2.6 \cdot 10^6$	$0.97 \cdot 10^6$	1,000.–	30 (oil) 90 (el)	12	8
TOTAL	--	--	--	$33.5 \cdot 10^6$	$10.3 \cdot 10^6$	3,236.–	787 –	24	4
G. Poolhouse	58			$281 \cdot 10^6$	$1 \cdot 10^6$	294.–	528 (oil) 360 (el)	300	4 months

spaced 21 cm apart, painted with flat black as collector
surface is positioned on top of a 2.5 cm (1") dense fiberglass
blanket. The width of each collector bin (between two sheets)
is 80 cm. There are 13 collector bins, 11 of them interconnected to heat the swimming pool, 2 of them to heat
domestic water, through heat exchangers (see Fig. 12).

Water is pumped onto the roof when a sensor on the roof
measures a temperature in the collector bin 5°C in excess of
the temperature of the pool. When the temperature drops below
this value, the water is drained (the system is open to air
via a vent at the reservoir).

For flow equalization and build-up of sufficient head
pressure, a regulating valve is installed near the top of each
bin.

Heat transfer to the pool is provided by pumping water
through the heat exchanger, heat transfer to the domestic
water tank is achieved by thermosiphon action, which enables
satisfactory stratification. The domestic H_2O solar storage
tank shown in Fig. 12 is connected in series with an oil-
fired domestic water tank.

This heating system was installed in the spring of 1976.
During the months of April through October, this system provided practically all of the necessary heat to keep the pool
at a temperature of 28°C (82°F). More than 50% of the pool
heat was supplied during February and November. However,
little was contributed during December and January due to the
shallow angle of the roof. Figure 13 shows the number of
hours of motor operation for the solar pump, pool and house
oil heating, illustrating this behaviour (pool heater oil
pump 4 l/min, house heater oil pump 8 l/min.)

House Heating

There is substantial overlap of solar pool heating and
house heat demand during February-April to justify installation of a wall (and ceiling) solar heating system in part of
the house. For this purpose, an 85 m^2 (946 ft^2) NW bedroom
wing was selected.

In this wing, the sheetrock at each outfacing wall was
removed, the space between the studs filled with insulating
pads, a 1.1 cm (1/2") Celutex layer added and a secondary
wall with offset studs installed (see Fig. 14). This wall
was covered with a 0.125 mm (5 mil) copper foil and, between
each stud, a 1 cm (3/8") vertical heating pipe connecting a
low feeder with a high collector pipe of 1.9 cm (3/4") O.D.

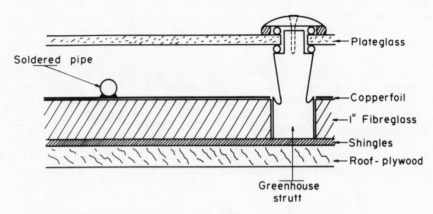

Fig. 11 Cross-section of roof collector built on
top of existing poolhouse roof.

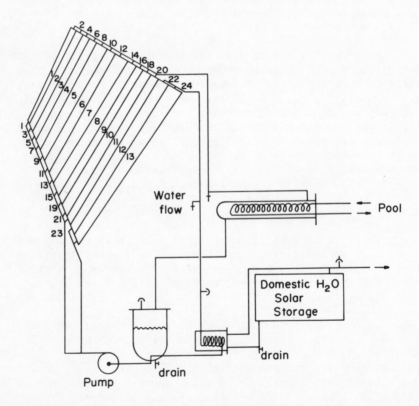

Fig. 12 Schematics of solar collector system

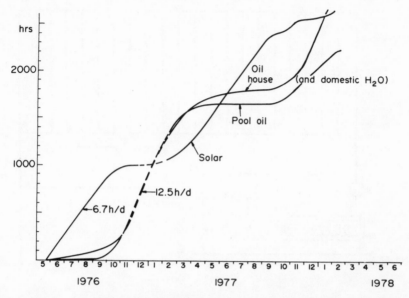

Fig. 13 Hours of operation of the pump motors for
solar and auxiliary equipment during the
first 22 months of operation.

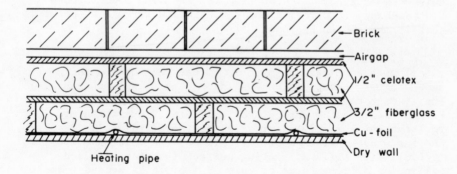

Fig. 14 Cross-section through wall in NW
bedroom wing

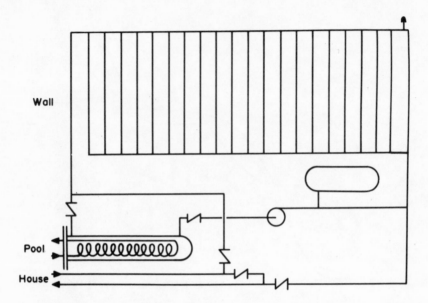

Fig. 15 Wall heating and cooling schematics

near the floor and ceiling respectively (see Fig. 15). The pipes were mechanically pressed against the copper foil. Then the wall was finished with a 1.1 cm (1/2") sheetrock, tape-joint, spackeled and painted, or wallpaper-finished.

Water flowing through pipes in walls and ceiling of this wing extracts its heat from the swimming pool through a heat exchanger. Circulation is controlled through a thermo-stat in the master bedroom.

All windows in this wing are double thermopane (four panes) with an airgap of 5 to 10 cm between the two thermo-pane windows, except for one sliding door (single thermopane). The average heating-season heat loss in this wing is estimated* as 2.8 Mkcal, using for outside walls (66.5 m^2) the heat resistance** R = 6(30), for ceilings (85 m^2 the value R = 7(35) and for windows/glass door (7.2/3.2 m^2) the value R = 2.8/1.25 (4/1.8). With 151.5 m^2 heating surface (outside wall and ceiling) a driving force of only $\Delta T \simeq 0.3°C$ is needed to make up for the heat losses through windows and door, and a total of 25,200 kcal is extracted from the pool through this system at a cold 25 DD (°C) winter day. This amount can be easily supplied by the pool oil heater in case of inclement weather; hence, this wing was completely separated from the previous heating system.

The system was completed in March 1977 and, except for an airlock in the ceiling in Dec./Jan. 77/8, the system operated highly satisfactorily.

An added benefit of the system is the equal temperature of nearly all room surfaces (except windows) which seems to have the effect of increased comfort and permits a slightly lower room temperature (~-2°C) compared to the other wing of the house with radiation ceiling heating when heating is adjusted to equal comfort.

House Cooling

The house was fully airconditioned. However, during the cooling season, the entire domestic water was diverted from the well directly through the wall (before entering the pressure tank) as indicated in Fig. 15, which provided an additional cooling means for the NW wing.

With an average use of 1000 l/day (this includes some use of water in the garden) at 12°C well head temperature, this presents an average cooling capacity of 10,000 kcal/day and permitted a substantial reduction of cooled air circulation from the airconditioning system into the NW wing (some use of this system is necessary to remove excessive humidity).

Performance Evaluation

The solar heating system is only partially installed. It is contemplated to add 80 m^2 collector surface on the SSW facing roof at an elevated angle of approximately 50° and wall heating in the remainder of the house (1 floor). This will permit substantial winter heating, however at the expense of excess heat production during the summer months, hence somewhat reduced overall utilization factor.

The present system utilizes almost all of its heat whenever produced (it provides little overheating of the covered pool, with its highest observed temperature at 32°), however, as indicated in Section 5, contributes little to the heating budget (pool and house) during November to February.

Table 3 summarizes the performance of the present system.

*The heat losses through the inside walls and floor are neglected since the neighbor rooms and the basement below this wing are kept at nearly the same temperature.
**Units of R are $m^2h°C/kcal$ and $(ft^2h°F/Btu)$ respectively.

Table 3. Energy Balance With and Without Solar Conversion for Certain Subsystems

	Heating Demand (Average)				Supplied by Oil				Supplied by Solar				Oil Saving kcal	Cost Saving $	% Solar Used
	Summer	11/1-2/10 kcal/day	10/15-4/15 kcal/day	year kcal	Summer	11/1-2/10 kcal/day	10/15-4/15 kcal/day	year kcal	Summer	11/1-2/10 kcal/day	10/15-4/15 kcal/day	year kcal			
Pool w/o cover	0.176	0.277	0.252	78	0.013	0.252	0.222	42	0.164	0.025	0.055	40	36	677	
Pool with cover	0.1	0.16	0.15	45	0	0.126	0.108	19	70.1	0.025	0.055	28	26	488	
Poolhouse	0.5/ season	15	15	15.5	0		15	15	0.5/ season			0.5	0.5	10	
Domestic Warm Water (est.)	0.05	0.076	0.076	23	0.013	0.068	0.05	11.5	0.038	0.008	0.025	11.5	11.5	216	
NW wing (est.)	$\dfrac{\text{cooling kWh}_{el}/y}{2000}$				$\dfrac{\text{cooling kWh}/y}{1000}$			2.8	$\dfrac{\text{cooling kWh}/y}{1000}$			1.4	1.4	60	26
TOTAL				86.3				46.9				39.4		800	46

The heating demand values for the covered and uncovered pool during mid-winter are obtained from the slope of the pool heating curve (Fig. 13) (comparing the 1976/7 with the 1977/8 season). The summer demand is estimated using the calculated heat loss through the pool cover. The poolhouse demand is calculated from the losses through walls, windows and ceiling.

The solar input to the pool is obtained from Fig. 13 for March-October and estimated from occasional calibration using a 24 point recorder to monitor the temperature distribution in the collection system and from the known flow rate through the collector (17 1/min through collector bins 1-11).

The domestic water heating was achieved by preheating the well water in the domestic water storage tank (Fig. 12) via thermosiphon from the heat exchanger, charged from bins 12 and 13 (bin flow rate 2 1/min). Up to July 1977, the house oil heater was left at its original setting and used an average of 1 hr/day, or 50,000 kcal/day oil heating even though sufficient solar energy input was provided to keep the upper stratus at the solar storage tanks above 60°C, except for a sequence of a few days of inclement weather. The water thermostat of the oil heater was consequently lowered and a valve installed to stop undesired back circulation through the furnace with the result that between July 15 and Sept. 15 only 17 hrs auxiliary oil heating was used, a reduction by a factor of almost 4.

The heating and airconditioning demand of the NW wing of the house was estimated from the design data (1,000 kcal/ DD (°C)). The solar heating contribution was estimated from the days in which the pool water was heated above 26°C (79°F) (below this temperature the water circulation pump for wall heating was not operated).

6. Performance Analysis

This analysis is of preliminary nature, since only part of the system is installed, and only a limited amount of data are collected (limited length of time of operation and limited instrumentation).

Therefore, more emphasis is given to overall performance data as obtained from the annual statistics of used oil and electricity. These data are then cross checked with the estimated performance of the components. Reasonable agreement indicates some justification for the used assumptions for the estimates of the component performance, as given in Section 5.

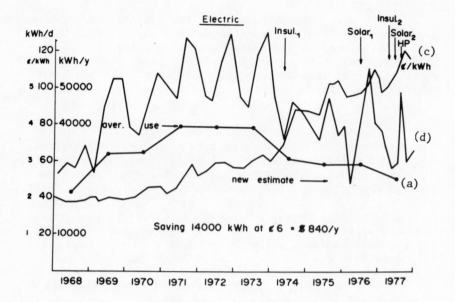

Fig. 16   Average daily (d) and annual (a) use of
           electric energy and cost (c) of electric
           energy from 1965 through 1977.

Energy Utilization Statistics

Figure 16 shows the daily use of electric energy as
obtained from the monthly bills during the years 1968-77.
The figure also contains the average yearly use and the
cost of electric energy for this residence. It is seen that
the use of electric energy has increased from 1968 to 1972.
This was caused by adding to the house the indoor swimming
pool (1969), the offices in the attic (1970/1) and finishing
the basement (1971).

A reduction of electric energy use in 1974 is caused by
better insulation (mainly reduction of airconditioning load
and electric office heating) and in 1976/7 due to installation of the poolcover (reduction of ventilation power).

Figure 17 gives a similar graph of the daily and averaged annual consumption of heating oil as well as the price
of the used oil.

Figure 18 shows a plot of the used oil per heating
season and a curve of the number of degree days, DD, per
heating season obtained from the University of Delaware's
heating plant at nearby Newark, DE. From these two curves,

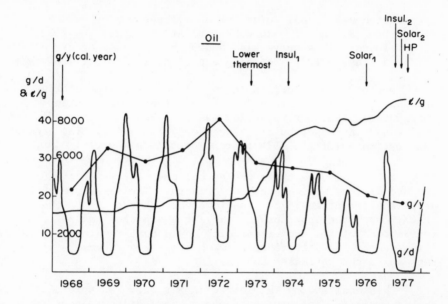

Fig. 17 Average daily (d) and annual (a) use of
oil in gallons (4 1), g, and price in
cents/gallon from 1968 through 1977.

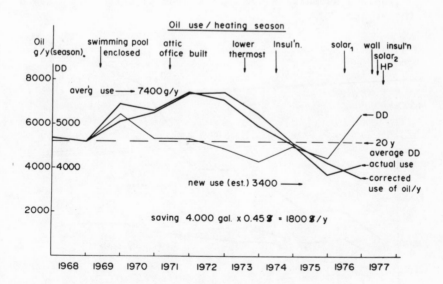

Fig. 18 Average seasonal use of heating oil
(corrected and uncorrected). See text.

a corrected use of oil curve for an average winter day (4,600 DD (°F) in DE., 5,000 DD (°F) at the house location is developed. This curve also shows the increase of use of oil from 1969 to 1971, reaching a plateau in 1972 of 7,400 g/y.

During the oil crisis in 1974, the thermostat setting was lowered from 23°C to 21°C with a marked reduction in consumption. Further reduction was achieved through insulation 1975/6 and introduction of solar energy in 1976/7. The present best estimate for the 1978 heating season is a use of 3,400 g/y.

In Fig. 19 are shown the actual cost curves of electricity and oil, as well as the sum of both. The sum curve shows a split starting in 1973. The lower branch represents the actual cost, the upper branch shows the expected cost if the consumption of oil and electricity would have remained at the 1973 level. The difference between these two curves represents the estimated cost savings.

This cost saving is plotted in Fig. 20. In here also, the accumulated cost saving is shown, indicating that up to the end of 1977 a total of ~ $8,000 has been saved by the incorporation of the conservation and solar conversion measures.

Payback Analysis

Table 4 summarizes the cost of the different components of the system.

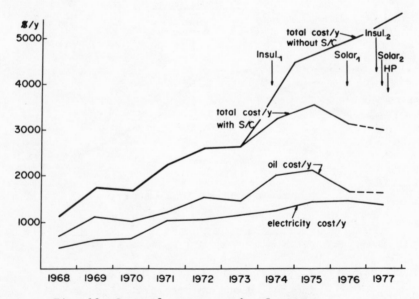

Fig. 19 Cost of energy used. See text.

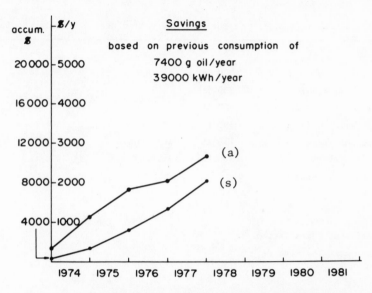

Fig. 20 Annual cost saving (a) and accumulated
saving (s).

For an economic analysis, the method of evaluating the
payback time is used (8).

Table 5 presents an estimate of the annual energy
saving (E), gives the cost/benefit ratio (E/C) and the
estimated payback time using an economical model with 5%
annual inflation, 6.5% interest rate and 9% estimated annual
increase of energy cost. For the insulation part, zero
maintenance costs are assumed, for the solar system, 2% per
year of the installation cost are assumed for maintenance.

It is seen that the systems original investment will be
paid back in approximately 6 years. Based on the actual
building time and energy savings as shown in Fig. 20, the
costs invested up to date are expected to be recovered by the
end of 1980.

This seems to be an encouraging result; however, it
should be pointed out that different components of the
system have quite different payback times. For further house
improvements, a careful analysis of such payback time will
determine the priority of the installation of such additional
items.

Table 4.   Cost Breakdown of Energy Conservation Means and Solar Equipment

INSULATION

Attic (blown in)	4200 ft^2	finished 7/29/74	(25¢/ft^2)	$ 1,044
Basmt. Ceiling Bats	800 ft^2	finished 7/29/74	(25¢/ft^2)	352
Garage Ceiling (bl/i)	600 ft^2			
Outside wall (bl/i)	990 ft^2	finished 2/22/77	(48¢/ft^2)	475
NW wing bal. insul	750 ft^2	finished 4/77	(27¢/ft^2)	200
Basement wall	200 ft^2	finished 4/77	(60¢/ft^2)	120
Ducts for oil furnaces				45
				$ 2,236

WINDOWS

3 gliders and installation			$ 550
Bedroom extra Thermopane	in NW wing finished 4/77		248
			$ 798

CURTAIN WALL

In living/dining room	320 ft^2	windows finished 7/77	($3.15/ft^2)	$ 1,000

POOLCOVER

Plastic foil		$ 27
Mechanics	finished 4/77	267
		$ 294

WALL HEATING/COOLING SYSTEM

Double wall incl. insul. and labor	750 ft^2	(60¢/ft^2)	$ 447
5 mil copper foil		(48¢/ft^2)	362
Plumbing in wall		($1.40/ft^2)	1,048
Related pipes, heat exch. el. parts	finished 4/77		1,127
			$ 2,984

SOLAR POOL HEATING SYSTEM (INCL. DOMESTIC WATER)

Roof collector incl. piping, heat exch., hot water storage for domestic water, pool-heat exchange, pump, insulation, electric connect. 720 ft^2 6/76		($12.12/ft^2)	$8,725

MAINTENANCE, REPAIR

Leak in pool heater	finished 2/77	$ 113
Replace plastic with copper pipes to collectors	finished 3/77	446
		$ 559
	TOTAL	$16,596

Table 5.   Payback Analysis of Energy Conservation Means and Solar
Conversion Equipment

	Total Cost C ($)	Savings ($/y)	Total Displ. Energy E ($)	E/C	Payback Time (y)
Insulation incl. windows and curtainwall	4,034	oil 437 el. 350	787	20%	5 years
Poolcover	294	oil 528 el. 360	888	302%	4 months
Solar pool and room heating	10,809	oil 740 el. 60	800	7.4%	13 years
TOTAL	15,137		2,475	16.4%	6 years

7.   Summary

The Solar One concept of simultaneous conversion of
sunlight into heat and electricity with rooftop deployment
of hybrid collectors and cost efficient utilization of these
forms of energy in conjunction with auxiliary supply from
conventional sources has been strengthened by experimental
results obtained from two installations, the Solar One
house of the University of Delaware and a retrofitted private
residence near Kennett Square, PA.

It is shown that in economically reasonable installation
air as heat transport fluid would require a minimum operation
temperature of the collector plate (solar cells) of 60° to
70°C for summer and winter operation respectively.

When water is used as a transport fluid, a substantially
lower operation temperature of typically 40°C can be maintained for the collector with a substantially increased
overall systems efficiency (typically 10–15%).

It is also indicated that low temperature heat can be
stored efficiently and utilized to heat a house satisfactorily.
In optimizing wall and ceiling heating by proper feeder tube
spacing, an attractive and cost efficient system may be
developed which provides a high degree of room comfort.

Finally, with a well insulated house, the heat storage
may be accomplished with a reasonable size storage tank: As

judged from the 85 m^2 NW wing with a heat demand of 1,000
kcal/DD (°C), a normal house with similar insulation may demand
2,000 kcal/CC (°C, hence for 35 DD (°C) storage and a reasonable tank storage $\Delta T \simeq 10°C$ for a very low temperature system,
the total storage volume needed is 7,000 1 (1,800 gal.)

However, more extensive experimentation is necessary of
a completed Solar One system to justify aggressive development
of such a combined system as one of the most promising solar
energy conversion systems.*

References

1. K. W. Böer, J. H. Higgens and J. K. O'Connor, Proc. IECEC
   Conference, Newark, DE.

2. H. M. Tan and W. W. S. Charters, Solar Energy 13, 121
   (1970)

3. V. Kolar, Int. J. Heat and Mass Transf. 8, 639 (1965).

4. T. M. Kuzay, M. A. S. Malik, S. M. Ridenauer and K. W.
   Böer, Solar One: First Results, Part 4, Univ. of Del.,
   Inst. of Energy Conv., Report (1974).

5. Solar One, Second Ann. Prog. Rep. to Power Utilities,
   Univ. of Del., Inst. of Energy Conv. (1975).

6. K. W. Böer, Proc. 9th IEEE Photovolt. Spec. Conf., Silver
   Spring, MD (1972) pg. 351 ff.

7. K. W. Böer, Chem. Tech. 3, 394 (1973).

8. K. W. Böer, Proc. Sharing the Sun, Intern. Conf. on Solar
   Energy, Winnipeg (1976) Pergamon Press, Vol. 9,
   page 1 ff.

9. D. G. Shueler and G. J. Jones, Proc. Am. Sec. ISES,
   Denver, Symposia Appendix (in print).

10. T. F. Halpin, R. Fischl, P. R. Herczfeld and D. B.
    Steward, Proc. Am. Sec. ISES, Denver, 2.2, 351 (1978).

11. K. W. Böer, Proc. ISES Conf., New Delhi, Pergamon Press,
    pg. 1593 (1978).

12. J. Duffie, 1978 Tutorials, Am. Sec. ISES, pg. 33 (1978).

---

*Obviously, the development of inexpensive solar cells to be
deployed in such hybrid collectors must be awaited.

COMMENTS

M.B. White (Princeton University): Would the author comment on the likely cost and performance of heating only the ceiling ? Many existing houses and even new houses do not lend themselves to wall heating because of windows, doors, closets, bookshelves, etc.   What would be the costs and performance of floor heating applied below the floor joists ?

Author's Reply: The cost of ceiling heating per unit area is slightly more expensive than wall heating because it requires plastering. Moreover, the copper foil cannot be installed in close contact with the copper pipes in the ceiling, hence there are larger temperature gradients in the ceiling than in the walls; consequently, the ceiling heating is less efficient. Heating in the floor from between floor joists has similar problems. The lowest ΔT can be achieved cost efficiently by wall heating.

J.T. Beard (University of Virginia): Would the author comment on the design decision to use copper foil rather than aluminum foil which is available on bat-type fibre glass insulation ?

Author's Reply: The copper foil in direct contact with copper pipes is more advantageous for reasons of avoiding corrosion. Aluminum foil on bat-type fibre glass insulation is too thin to be an effective fin for the copper pipes 18 inches apart.